T0315368

GREGOR MENDEL

His Life and Legacy

DANIEL J. FAIRBANKS

Guilford, Connecticut

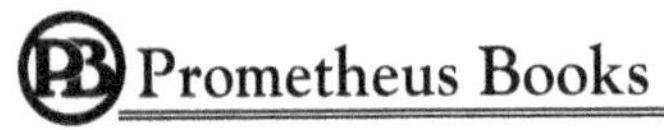

An imprint of Globe Pequot, the trade division of
The Rowman & Littlefield Publishing Group, Inc.
4501 Forbes Boulevard, Suite 200, Lanham, Maryland 20706
www.rowman.com

Distributed by NATIONAL BOOK NETWORK

British Library Cataloguing in Publication Information Available

Library of Congress Cataloging-in-Publication Data

Names: Fairbanks, Daniel J., author.
Title: Gregor Mendel : his life and legacy / Daniel J. Fairbanks.
Description: Lanham, MD : Prometheus, [2022] | Includes bibliographical references and index. | Summary: "This biography of Gregor Mendel, the founder of genetics and one of the most ingenious and influential scientists in history, is written especially for those whose background in science may be limited"—Provided by publisher.
Identifiers: LCCN 2021054683 (print) | LCCN 2021054684 (ebook) | ISBN 9781633888388 (cloth) | ISBN 9781633888395 (epub)
Subjects: LCSH: Mendel, Gregor, 1822–1884. | Geneticists—Austria—Biography.
Classification: LCC QH31.M45 F35 2022 (print) | LCC QH31.M45 (ebook) | DDC 576.5092 [B]—dc23/eng/20211216
LC record available at https://lccn.loc.gov/2021054683
LC ebook record available at https://lccn.loc.gov/2021054684

∞™ The paper used in this publication meets the minimum requirements of American National Standard for Information Sciences—Permanence of Paper for Printed Library Materials, ANSI/NISO Z39.48-1992.

CONTENTS

Preface

THEN WHAT COULD DEATH DO *if thou shouldst depart, leaving thee living in posterity?* This line from Shakespeare's sixth sonnet epitomizes people's perception of heredity. Distinct characteristics in parents reappear in their offspring, "living in posterity" from one generation to the next. The resemblance of offspring to their parents is readily observable in plants, animals, and humans. Yet for most of human history, no one knew quite how inheritance works. Over the course of millennia, farmers bred animals and plants to be more productive and desirable, choosing those with the best inherited characteristics as parents for each subsequent generation. Countless people dreaded so-called family curses: serious medical disorders that appeared through multiple generations in extended families in some but not all individuals, sometimes in every generation, sometimes skipping generations. Despite innumerable observations of heredity, no one noticed how it consistently portrays an elegant mathematical pattern, hiding in plain sight. The story of its discovery, neglect, and then rediscovery is one of the most intriguing in the history of science, tracing its origin to a nineteenth-century Augustinian friar named Gregor Mendel.

For me, the discovery of Mendel's genius was an epiphany. As an undergraduate university student, I was trying to decide whether to follow a four-generation tradition in my family of professional artists or another family tradition of medicine. I had enrolled in a university course in genetics, part of the required curriculum for premedical education. My professor was a consummate scientist, dedicated to his students and his scientific research, his name James Farmer. I can still hear in my mind his deep baritone voice permeating the classroom of 250 students, captivating my attention in every lecture. The first topic of the class was Mendel's experiments and how the theory he inferred from them founded the science of genetics. As students, we scrutinized his experiments in detail, exploring how he planned their design and how he analyzed the epic numerical data he collected for eight years to develop what is now known as Mendelian genetics, the fundamental principles of inheritance.

Mendel's experiments were to me like a masterwork of fine art—logical, symmetrical, and exquisitely beautiful. We were not required to read Mendel's article, but I did so on my own. I could almost hear Mendel's voice as a teacher, telling his students beforehand the patterns to seek, providing the numerical evidence, and then reiterating those patterns to show how they emerged from the numbers. I knew at that moment that I wanted to be a geneticist—and a teacher.

Dr. Farmer somehow perceived my fascination and invited me to be his research assistant. For two years, I spent hours in his laboratory, observing thousands of fruit flies I had anesthetized with ether, allowing them to sleep safely for about fifteen minutes on a pane of glass as I sorted them with a paintbrush under a dissecting microscope according to their different eye colors. I tabulated the numbers, sometimes in the hundreds, before returning them to the bottles where we kept them. Among those flies, patterns resembling those Mendel observed in his pea experiments emerged quite literally before my eyes as I discovered and documented scientific phenomena that no one had previously observed. I soon found that discovering something new and meaningful about nature is what drives scientists to be so passionate about their work. My first publication came from that undergraduate-student research.[1]

I began voraciously reading books and articles on genetics. In doing so, I realized the motivational power of biographies. I read three that were especially influential. One was by Evelyn Fox Keller, *A Feeling for the Organism*, an appropriately titled biography of Barbara McClintock, whose experimental genius and extraordinary observations led her to deduce how segments of DNA could move within genomes, her work based on the foundation Mendel had laid.[2] She received the 1983 Nobel Prize in Physiology or Medicine because her discoveries in the corn plant applied to humans, just as Mendel's discoveries in the pea plant also applied to humans. Another book I read carries the title *Facing Starvation* by Lennard Bickel. It is a biography of Norman Borlaug, recipient of the 1970 Nobel Peace Prize for his applications of Mendelian genetics in wheat, leading to astounding increases in food production in impoverished parts of the world, now known as the green revolution.[3]

I chose to pursue plant genetics in graduate school, inspired in large part by these biographies of McClintock and Borlaug. I later was fortunate to spend one-on-one time with each of them while I was a graduate student. Although I had chosen genetics as my career, I had not abandoned my ambitions as an artist. The time I spent with Borlaug was in my studio as he posed for a commissioned portrait bust I made of him. That bronze statue is now on exhibit in Borlaug Hall on the University of Minnesota campus and another bronze casting at the Centro de Investigaciones Agrícolas del Noroeste (CIANO), a scientific research institute in Mexico where Borlaug conducted much of his innovative research in wheat.

I have continued my career as an artist. The book you are now reading contains several of my illustrations of people, places, and topics important to Mendel, some of them held in the permanent collection of the Mendelianum of the Moravian Museum in Brno, the city where Mendel lived for most of his life. It is a lovely city, the historic capital of the Moravian region in what is now the Czech Republic.

The other influential biography I read was on Mendel. The author, Hugo Iltis, was a high school teacher in the early twentieth century in Brno. Several of the people who knew Mendel were still alive at the time, and Iltis personally interviewed as many as he could. Others sent him letters with their reminiscences. He assembled documents found at the monastery where Mendel lived, at the University of Vienna where he studied, in the school buildings and institutes where he taught and worked, and in the homes of his family members and friends. Iltis wrote the book in his native German, published in 1924.[4] The version I read is an English translation published in 1966.[5]

Inspired by Iltis's biography, I visited the place where Mendel lived and conducted his famous experiments. It felt like a pilgrimage, largely because Mendel's life and contributions had become so meaningful to me but also enhanced by the fact that the site was an old fourteenth-century monastery. As my colleague, professor and artist Marcus Vincent, and I arrived, Anna Matalová greeted us at the door. She was head of the Mendelianum, a museum and archive dedicated to Mendel. Although we had never met before that day, she promptly set aside her work to lead us on a personal tour. I had read in Iltis's biography that Mendel owned books by Charles Darwin and had made annotations in them. At my request, Dr. Matalová retrieved several of those books from the vault, including his personal copy of Darwin's *Origin of Species*. With her permission, I photographed Mendel's annotations in those books. This was the beginning of what turned out to be a productive research career on Mendel, encouraged and facilitated by Dr. Matalová for the past twenty-eight years. She is one of the finest, most gracious, and most supportive scholars I have met. I unhesitatingly dedicate this book to her.

Why write a new book on Mendel? Importantly, it is unique in several ways. It is much more concise in scope and detail than the three standard scholarly biographies, all of which are exceptional. The first of these is the one by Iltis I mentioned a moment ago. The second is a superb book titled *Gregor Mendel: The First Geneticist* by Vítězslav Orel, an English translation of Orel's original biography in Czech, published in 1996.[6] Orel was one of the world's foremost scholars on Mendel and head of the Mendelianum who preceded Dr. Matalová in that role. The book, unfortunately, is no longer in print. The third is an exhaustively researched and detailed two-volume work titled *Solitude of a Humble Genius* by Jan Klein and Norman Klein. The first volume, consisting of Mendel's early years, was published in 2013.[7] At the time of this writing, the second volume has yet to be published but is highly anticipated by those of us who study Mendel's history. I have relied heavily on all three of these extensive biographies while writing this book and express my unbounded gratitude to their authors, two of whom, Iltis and Orel, are now deceased. Two other books have been especially helpful: *Origins of Mendelism* by Robert Olby[8] and *The Origin of Genetics: A Mendel Source Book* by Curt Stern and Eva Sherwood.[9] Several popular biographies are also available, some as stand-alone books, others as short sections of more expansive books, and a few as books for children. Countless historical articles and brief biographies of Mendel are likewise available in a wide range of formats and languages—and a wide range of accuracy.

The book you are now reading is considerably shorter than the standard biographies, readable in a few hours. Its intended audience is casual readers, especially those who may have limited scientific backgrounds. It adheres to the high standards of scientific and historical accuracy evident in the standard biographies, an aspect that is especially important given the extent of inaccuracy and embellishment that is unfortunately prevalent in many accounts related to Mendel, a tendency Orel referred to as "Mendelian mythmaking."[10]

Another aspect of this book is its inclusion of new information on Mendel's history that continues to emerge, fueled by a powerful resurgence of research in recent years. I am honored to be immersed in that research. In this book, you will read new perspectives founded in my own studies and those of exceptional scientists and historians who willingly work with me as we jointly share our discoveries and insights. Several of them have coauthored articles and books with me. To all of them, I extend my gratitude. Most notably, I wish to acknowledge the assistance of Scott Abbott with German–English translation and the graciousness of Anna Matalová, Eva Matalová, and Jiří Sekerák, the current head of the Mendelianum, for reviewing the manuscript.

I also thank the editors and staff at Prometheus Books and Rowman & Littlefield for their exceptional professionalism, expertise, and assistance. Jonathan Kurtz, executive editor of Prometheus Books, has been especially encouraging and passionate in support of this book. His efforts have truly made it possible. I am also grateful to Nicole Carty and Bruce Owens for their exceptional editing.

Finally, the timing of this book celebrates a magnificent event. Mendel was born in July 1822 (the day of his birth is disputed, as you will soon read). The year 2022, therefore, is the bicentennial of his birth, with events scheduled throughout the year. This book is one of several publications commemorating this bicentennial.

Daniel J. Fairbanks
September 2021

Prologue
A Misty Winter Evening

A THICK FOG SETTLED IN the cold and still air of Moravia's largest city, nestled among gentle hills in the geographic center of the European continent. The day was Wednesday, January 11, 1865, the high temperature barely above freezing. By early afternoon, according to the local weather report, *Sonnenschein* (sunshine) made a fleeting appearance, only to disappear as the heavy mist returned shortly thereafter.[1] As darkness fell, several dozen men carefully traversed their varied paths through the icy cobbled streets, relying on freshly kindled gas lamps as guiding beacons. They arrived one by one from various directions at a prominent building housing a school known as the *Realschule*. Rows of arched windows adorned the four-story exterior, its Italian-style architecture reminiscent of Florence's Medici Palace (figure P1). As the time arrived, bronze bells tolled within the domed campanile, their resonance permeating the shrouded air to mark the hour.

The attendees were of two cultures. Some had Czech surnames, others German. Their city, situated on the main thoroughfare between Prague and Vienna, likewise bore two names reflecting the two principal ethnicities of its inhabitants. Those who spoke Czech called it Brno, the first three consonants squeezed into a syllable with no vowel to separate them, the terminal "o" elongated. Its German name was Brünn, the guttural "r" and umlauted "ü" dominating the pronunciation with a gravelly sound. The meeting would be in German, the principal language of the Austrian Empire.

The city bustled with powerful nineteenth-century industrial expansion, its growth extending well beyond the remnants of the old town walls. A towering Gothic cathedral, bearing the names of Saints Peter and Paul, crowned a hill near the old-town center, overseeing onion-dome steeples

Figure P1. The Realschule, a school where Gregor Mendel and Alexander Makowsky taught science and where the Natural Science Society convened for its monthly meetings. *Drawing by Daniel J. Fairbanks from sketches and photographs on-site, sepia pastel on paper. Collection of the Mendelianum, Moravian Museum, Brno.*

punctuating the skyline throughout the city. Their collective presence proclaimed the city's predominant faith as Roman Catholicism. A massive medieval fortress topped an even higher hill, known alternatively in the two languages as the Špilberk or Spielberg Castle. With austere walls faced in pale-yellow stucco, and barred windows reminiscent of its time as one of Europe's most dreaded prisons, it cast an ominous presence over the surrounding countryside.

The men on that cold and foggy evening gathered for the monthly meeting of the Natural Science Society in Brünn. The organization was still young, at the beginning of its fourth year. Even so, it was astoundingly popular with almost 300 names on its membership roster. Accompanying each name was the individual's profession—schoolteachers, professors, physicians, engineers, pharmacists, bankers, attorneys, accountants, government officials, military officers, innkeepers, clergy, and manufacturers of a wide range of products. All had a passion for science, some by profession, others by avocation or curiosity.

The featured speaker that January evening was one of the society's founding members, Alexander Makowsky, a gentleman with straight dark hair, slicked and combed back, his bushy moustache merging into equally bushy sideburns. He commanded high esteem as an accomplished botanist, geologist, and science teacher at the *Realschule*. His chosen topic was sweeping Europe like an intellectual tsunami, bound to elicit vigorous discussion—Charles Darwin's book *On the Origin of Species*.

As a skilled teacher, Makowsky knew how to captivate an audience. His words were passionate as he advocated for Darwin's book, quoting well-chosen passages from a recent German translation. "These, *meine Herren!*" he emphatically concluded, "are generally the main features of Darwinian theory, whose difficulties, however, the author himself could not conceal. As much as it may stand in contradiction to past views, it has at least the same legitimacy as the contrary view that species are immutable products of nature. It makes natural science no less impossible, than did astronomy, when it was discovered that the earth moves."[2]

Sitting in the audience was a Roman Catholic priest—his name Gregor Mendel. He too was a founding member of the society. Makowsky was his close friend and scientific collaborator. Twenty-one years earlier, Mendel had joined the Augustinian Order of St. Thomas located in a monastery on the western outskirts of town. The members of his order were friars, not monks, dedicating themselves to secular learning, scholarship, and teaching, especially in philosophy and the sciences. Several were highly educated and internationally renowned. Like his friend Makowsky, Mendel was a science teacher at the *Realschule* (figure P2).

The friars of the St. Thomas Monastery shunned the cloistered devotional environment of monastic hermits, preferring academic research, teaching, and service to the people. They were visible and admired not only in the city but throughout the surrounding region as well. They served as professors and administrators at schools and universities, board members of prominent businesses and organizations, overseers of the monastery's large agricultural holdings, and participants in scholarly societies, as was the case for Mendel this evening.

Makowsky's presentation was familiar territory for Mendel, who was an experienced physicist, mathematician, botanist, and science teacher. He had studied at the University of Vienna under Professor Franz Unger,

Figure P2. Members of the Realschule faculty, Alexander Makowsky standing left of center, Gregor Mendel seated on the right. This photo was taken during the 1864–1865 academic year, the same year when Mendel and Makowsky presented their lectures to the Natural Science Society. *Period photograph, collection of the Mendelianum, Moravian Museum, Brno, Czech Republic, public domain.*

a world-renowned botanist and paleontologist who popularized organic evolution in a series of newspaper articles and in a magnificent book with hand-tinted lithographs depicting how fossilized plants and animals appeared when they lived eons ago during the earth's ancient geologic periods. Under Unger's tutelage, Mendel had studied biological evolution seven years before Darwin's *Origin of Species* was published. When the book appeared in German translation, Mendel acquired his own copy and read it, annotating passages he found relevant to his work. Makowsky quoted from this same translation during his speech, highlighting several of the same topics Mendel had marked.[3]

The dichotomy pitting evolution against the traditional view that all species were fixed from the time of creation had come painfully close to Mendel. In Vienna, he witnessed firsthand a series of vitriolic attacks by Sebastian Brunner against the evolutionary teachings of his professors. Brunner was a prominent Catholic priest, born and schooled in Vienna, a prolific author, professor of theology, poet, witty satirist, and gifted orator—also an anti-Semite and vocal defender of strict Catholic orthodoxy.

Brunner had appointed himself as a modern Ezekiel, a watchman warning Vienna's citizenry of the twin evils of secular science and pantheism, which he believed were corrupting his native city. In a newspaper headline, he christened Professor Unger as "Isis Priest and Philistine," sarcastically declaring that Vienna's botanists "do everything they can to make themselves into plants of botanical learning that can be smelt from afar—and place themselves voluntarily into the eternally stinking dung-bed of the pantheistic world view, which nevertheless fosters a certain richness of blossoms."[4]

Through much effort, Mendel had established himself as one of those Viennese botanists. As his time at the University of Vienna was drawing to a close, Mendel's professors and colleagues elected him as a full member of the prestigious Imperial-Royal Zoological-Botanical Society in Vienna. He presented his research to its membership in the monthly meetings and published his first two scientific articles in the annals of its journal.

As an avowed and freethinking scientist, Mendel could not evade official scrutiny of his intellectual pursuits. News of increasing intellectualism and secularization in Austrian monasteries had reached the Vatican. Pope Pius IX charged Prince Friedrich Johann Jacob Celestin von Schwarzenberg, cardinal of the Holy Roman Church and archbishop of Prague, to carry out "visitations" at Austrian monasteries to purge them of humanistic philosophy and science. The cardinal assigned the task for the St. Thomas Monastery to the bishop of Brünn, Anton Ernst von Schaffgotsch, a staunch purveyor of orthodoxy who was already suspicious of Mendel and his erudite brethren.

The visitation was, in fact, the culmination of a lengthy investigation dragging on for more than a year. In the end, Schaffgotsch issued a scathing report, penned in Latin, insisting that the St. Thomas Monastery be disbanded. His missive declared, "In the house tending to the Rule of Saint Augustine reigns a secular spirit which the few lappets of the Augustinian habit fail to cover up," which had extinguished "the last ray of spiritual life." The bishop singled out Mendel as one who "studies profane sciences at a worldly establishment in Vienna at the expense of the monastery to become a professor of said sciences at a state institution."[5] Cardinal Schwarzenberg endorsed the bishop's demand for dissolution the community of friars to which Mendel belonged and delivered it to the Vatican for final approval. It languished there for years, interminably mired in administrative limbo.

As the year 1865 opened, the perception that science posed a threat to church authority had escalated well beyond the Austrian Empire. Barely a month before Makowsky's lecture on Darwin, Pope Pius IX issued from the Vatican his famed encyclical, the "Syllabus of Errors," a list of eighty heresies that threatened Christianity, among them the infusion of rationalism, naturalism, pantheism, and liberalism into science.

The rising popularity of Darwin's theory was a principal undercurrent prompting the pope's encyclical. Troubled by the threat the encyclical posed to science, Thomas Henry Huxley, Darwin's trusted friend and colleague, hastily penned a rebuke infused with moral outrage, published on the last day of the year in 1864. He concluded his exposé with praise for the growing body of scientists: "Since the world began, there never has been so deep a reverence for truth, . . . so earnest a desire to build up some theory of this wonderful universe . . . as among the scientific workers of this age and generation."[6]

Against this backdrop, Mendel would be the next featured speaker in the Brünn Natural Science Society's monthly meetings—a devout priest speaking on science. His presentation was ambitious, so extensive that it occupied two full monthly sessions: February 8 and March 8, 1865. In the first of these two lectures, Mendel presented an account of meticulously designed experiments he had conducted over a period of eight years on an epic scale. In the second, he proposed a sweeping theory synthesized from the results. His experimental organism was the common garden pea. His approach flowed with mathematical sophistication, an analysis of exhaustive numerical data to buttress his theoretical inferences. The topic logically followed Makowsky's January presentation, clarifying the role of hybridization and heredity in the evolution of new species, a topic Darwin had addressed at length in the eighth chapter of *Origin of Species*, the chapter Mendel annotated more than any other in his personal copy of the book.

Based on the data he collected, Mendel explained how fertilization must operate at the level of cells, surmising that two cells must unite, one from the female parent and one from the male, each contributing hereditary elements equally to the offspring. This was a theoretical leap at the time. Mendel's professor, Franz Unger, had postulated such parental equality, but others disputed it. Among the detractors was Mendel's other botany professor at the University of Vienna, Eduard Fenzl, who argued that heredity was purely paternal, the female serving simply as a nurse to the heredity-bearing pollen.

Never had such a massive amount of data as Mendel's definitively confirmed the hypothesis of parental equality and refuted the notion of strict paternal inheritance. Mendel's conclusion ultimately proved accurate not only for plants but also for animals and humans: one egg cell unites with one sperm cell at fertilization, both contributing to the offspring.

This inference, based on abundant evidence, was a major discovery. But it was only the beginning of a sweeping theory Mendel presented to his colleagues. Two other monumental achievements emerged from his work, constituting a framework that would ultimately reveal the continuity of life. The first was his deduction that underlying material units conferred inherited traits. He named these units "formative elements" (*bildungsfähige Elemente*) and emphasized their "material nature" (*materielle Beschaffenheit*). We now call these formative elements *genes*, their material nature *DNA*. The second was his mathematical extrapolation of the fundamental principles of heredity—the laws that govern how offspring inherit those elements from their parents, generation after generation.

No one, including Mendel, realized at the time the magnitude of his discovery. Thirty-five years later, at the dawn of the twentieth century, his experiments and his theory would be rediscovered—sixteen years after his death. Soon thereafter, his growing number of disciples, self-named as Mendelians, employed it to establish a science so novel that it required a new name: *genetics*.

It is often said that Mendel's lectures fell on deaf ears. One author went so far as to characterize his lectures as "the throwing of pearls before swine."[7] Another claimed,

> Stolidly the audience had listened. Just as stolidly it has risen and dispersed down the cold, moonlit streets of Brünn. No one had ventured a question, not a single heartbeat had quickened. In the little schoolroom one of the greatest scientific discoveries of the nineteenth century had just been enunciated by a professional teacher with an elaborate array of evidence. Not a solitary soul had understood him.[8]

Contemporary accounts tell a less dramatic story. In a newspaper article recounting the February presentation, the author reported, "That the theme of the lecture was well chosen and the exposition of it entirely satisfactory was shown by the lively participation of the audience."[9] The March report was similar, recounting how Mendel explained the role of cells in fertilization and that his lecture was "very well received."[10] In another newspaper article consisting of the minutes of the March lecture, the author astutely noted that Mendel's "large number of ingenious experiments, supported by the best achievements, should contribute not a little to the explanation of this process, which has hitherto been imprecisely observed."[11]

Mendel himself was less optimistic. "I knew the results I had obtained were not easily compatible with our contemporary scientific knowledge," he wrote to a colleague, and "I encountered, as was to be expected, divided opinion."[12] The text of Mendel's lectures appeared in print in 1866 in the society's journal, an article that is now a classic in the history of science. Its value is not only scientific. In 2019, a rare original reprint, one of forty delivered to Mendel, commanded 287,250 British pounds by auction, the equivalent of approximately 357,000 U.S. dollars.[13]

Neither Mendel nor any of his contemporaries realized how revolutionary his theory was. Unlike Darwin, who witnessed his theory achieve immediate worldwide fame (and infamy), Mendel would never know how powerfully his theory would impact science—and humanity. "The laws governing inheritance are quite unknown," Darwin lamented just a few pages into the first chapter of *Origin of Species*.[14] By 1865, Mendel had discovered and presented those laws, which ultimately would bridge the most gaping chasm in Darwin's theory—belatedly years after both Mendel and Darwin had died.

In a twist of coincidence, Darwin devised his own ill-fated hypothesis of inheritance that same year, 1865, which he named with words conveying his own uncertainty: "the provisional hypothesis of pangenesis." A few years later, Mendel learned of Darwin's hypothesis when he read a German translation of the book wherein Darwin proposed it. Mendel's copy of the book contains his handwritten note of reserved cynicism: *sich einem Eindrucke ohne Reflexion hingeben*, "to indulge in an impression without reflection."[15]

When Mendelism and Darwinism finally converged in the twentieth century, the unified theory that emerged would be hailed as the "modern synthesis," one of the most momentous revolutions in science. Not only was this synthesis a theoretical tour de force, but it also spurred countless advances to benefit human health, world food production, and the global economy. Nearly everyone on the planet has immensely benefited from these advances, most unknowingly.

Through the years, Mendel's experiments and his theory generated controversies that have risen and fallen. Early in the twentieth century, a group of influential scholars elevated his experiments and his theory to the highest echelon of scientific achievement; others pilloried Mendel with accusations of fraud.[16] Throughout modern history, proponents and detractors alike have interwoven Mendel's theory into their own worldviews, fueling the flames of disputes and prolonging political battles. A few courageous scientists under Stalin's dictatorial regime were imprisoned, several choosing to sacrifice their lives rather than renounce their acceptance of Mendelism. As their leader, Nicolai Vavilov, proclaimed, "We shall go to the pyre, we shall burn, but we shall not retreat from our convictions."[17] Vavilov was true to his word, paying the price of those convictions with his life when he died of malnutrition as a political prisoner. Mendel's theory has served as the foundation for some of history's greatest scientific advancements while alternatively being improperly co-opted to justify some of the world's most horrendous atrocities.

This book commemorates Mendel's 200th birthday (July 22, 1822), highlighting his life, his groundbreaking discoveries, and the theory of the gene he synthesized from them—a theory that has endured as one of the most powerful achievements in science. His contemporaries often bear posthumous ridicule for their failure to grasp the magnitude of his discovery and its pioneering implications. Yet they can hardly be blamed. Despite the eloquence and lucidity of Mendel's presentation, even modern readers struggle to understand it. A key purpose of this book is to interpret Mendel's discovery and his theory in a manner that makes sense to casual readers, especially those whose background in science may be limited. The book also untangles some of the many myths that shroud Mendel, his history, and his scientific accomplishments, setting them in the context of his time and revealing who he was, what he discovered, and what he really said.

From Poverty to Priesthood 1

To dispel the dark power of superstition that weighs heavy on the earth

—JOHANN MENDEL

THE SERENITY OF THE SILESIAN FARMLANDS in the early nineteenth century, straddling what is now the border between the Czech Republic and Poland, belied a troubled region pushed and pulled by empires on all sides. The land and its people had at various times been under Hungarian, Germanic, Prussian, and Austrian rule. The Napoleonic Wars, which dragged much of central Europe into seemingly unending bloodshed, had concluded a mere seven years earlier. Now, in 1822, a negotiated border divided the region into Prussian Silesia on the north and Austrian Silesia to the south. A dialectic mixture of languages permeated the streets of its cities and towns. A small ethnic-German town, known at the time as Heinzendorf, hugged the southern border of Austrian Silesia, the much larger region of Moravia to the south. Today, the town is in the Czech Republic and carries the name Hynčice.[1] It was then and is now an elongated group of homes, all close to a central road paralleling a stream, with cultivated fields bordering the settled area.

Rosina and Anton Mendel resided in one of those homes, number 58 on the town's map, where they made a meager yet sustainable living (figure 1.1). Anton had been a soldier in the Napoleonic Wars, although little is known of his military experience other than his early discharge in 1816 after Napoleon's defeat. Rosina and Anton married in 1818, welcoming their first child in 1820, a daughter whom they named Veronika. In July

1822, Rosina gave birth to a boy, his chosen name Johann. The exact date of his birth has been disputed as either July 20 or 22. Records in the village church list his birthday and baptism as July 20, along with birth and baptismal records for several other infants born in 1822. Record keeping at the time was imprecise, with little certainty to identify whether or not each child was actually born on the day listed in the records. Mendel himself consistently claimed July 22 as his birth date, the day when he and his family celebrated it. That same day was also meaningful as the *Magdalenentag*, the Feastday of St. Mary Magdalene.[2]

Figure 1.1. Mendel's childhood home in Heinzendorf near what was then the southern border of Austrian Silesia. *Drawing by Daniel J. Fairbanks from sketches, a painting, and photographs, sepia ink and pastel on paper. Collection of the artist.*

Three of Rosina and Anton's children lived to adulthood: Veronika and Johann were older, and Theresia, born in 1829, was nine years younger than her sister and seven years younger than her brother (figure 1.2). Two daughters died early in life. The first, named Rosina after her mother, was born in 1825 and died as a toddler in 1828. Another daughter, also named Rosina after her mother and deceased sister, was born in 1831. She died just thirty-three days after her birth.[3] Veronika and Johann were old enough to comprehend the loss of their two sisters and undoubtedly suffered terrible grief.

Figure 1.2. Gregor Johann Mendel and his two surviving sisters in order of age as they appeared in adulthood, Veronika Mendel Sturm (left), Gregor Johann Mendel (center), and Theresia Mendel Schindler (right). *Drawings by Daniel J. Fairbanks from historic photographs, sepia pastel on paper. Collection of the artist.*

The family was Roman Catholic, as were most of the villagers in the surrounding area. Although Protestantism (mostly Lutheran) and Judaism had been allowed to expand in the Austrian Empire, Roman Catholicism still reigned as the dominant religion. And it would dominate Johann's life. He and his family attended services in the small Church of Saints Peter and Paul in Groß Petersdorf (Vražné), a village that borders Heinzendorf (figure 1.3). Twenty-one years later, he would assume the monastic name Gregor when he joined the Augustinian Order of St. Thomas.

Figure 1.3. The Church of Saints Peter and Paul in Groß Petersdorf (Vražné), a short distance from Mendel's childhood home. *Drawing by Daniel J. Fairbanks from sketches and photographs, sepia ink and pastel on paper. Collection of the artist.*

Farming, raising animals, beekeeping, gardening, and fruit-tree cultivation captivated much of Johann's childhood, as these were his family's means for providing food and making a living. He would pursue these interests throughout his life, tending gardens and beehives and managing the agricultural holdings of his monastery when he became its abbot. Johann's mother's sister had married his father's brother, and they lived nearby, sharing the farmland and duties. Consequently, Johann and his sisters were raised alongside their double-first cousins.

Elementary education was free and compulsory, combined with religious instruction. Johann and his sisters attended the local school where they studied under two teachers, Pfarrer Johann Schreiber, who was also the local priest, and Thomas Makitta, a certified teacher and by then an elderly man who had been at the school for decades, teaching not only the Mendel children but also their parents when they were children.

As the only son in the family, Johann faced a dilemma from the days of his childhood. Should he remain on the family farm, eventually taking it over, or pursue his education away from home? His teachers recognized how precocious the eleven-year-old boy was, and they persuaded his parents that he should enroll for a year at a *Hauptschule*, a boarding school for intellectually gifted students. The school had an excellent reputation, accepting only the most accomplished pupils on the recommendation of their teachers. Two other students from Heinzendorf had attended and done well. There was every expectation that young Johann would excel there also.

The decision challenged his parents. Boarding school came at a high price, a financial sacrifice that most rural families could not afford. The school was in the Piarist Monastery in nearby Leipnik (Lipník), and tuition was free, but the family struggled to assemble the resources needed to support his room and board.

The monastery consists of a large rectangular set of connected buildings with two interior courtyards, attached to the Church of St. Francis of Assisi, its tall and pointed belfry in Italian style dominating the skyline of the town center (figure 1.4). Here Johann first became acquainted with Czech culture and language. The Silesian dialect of German was the language of his home village. Yet in Leipnik, only a third of the citizens spoke German; the majority spoke Czech.[4]

Johann performed exceptionally well, with grades listed as *eminens* (Latin for "eminent") and a side notation in German, *erster Prämiant seiner Klasse*, "first prize for his class."[5] His performance qualified him to enter the Gymnasium in Troppau (Opava), where he enrolled in December 1834 at

Figure 1.4. The Piarist Monastery and Church of St. Francis of Assisi in Leipnik (Lipník), where Mendel attended boarding school in 1833–1834. *Drawing by Daniel J. Fairbanks from a historic photograph, sepia ink and pastel on paper. Collection of the artist.*

the age of twelve. The school's headmaster was Ferdinand Schaumann, a friar from the Augustinian Order of St. Thomas in Brünn, who had been assigned to the post in Troppau. Such assignments for members of the St. Thomas Monastery were common. Johann would eventually join the same order and spend much of his career as a schoolteacher. At the Troppau Gymnasium, he consistently scored among the top students, his designation *prima classis cum eminentia*, "first class with eminence."[6]

Troppau was the largest city of Austrian Silesia, a cosmopolitan town consisting of numerous cultures, lying just south of the border between the Austrian and Prussian empires. The majority of its inhabitants spoke Johann's native dialect of Silesian German. Nonetheless, on its streets, he heard Prussian Silesian, similar to Polish, and also Sorbian and Ashkenazi Yiddish.

The Gymnasium occupied part of a large set of buildings, three stories high, surrounding a square courtyard and attached to the Church of Saint Adalbert with an Italian facade, Corinthian columns, and statuary of the saints (figure 1.5). Its organ was the centerpiece of its interior baroque architecture, famed throughout Europe for its musical grandeur. Also studying here at the same time as Johann was a boy named Karl Krischkowsky, a musician who would later become one of Mendel's closest friends and colleagues as a fellow friar in the St. Thomas Monastery.

Johann studied at the Gymnasium for the next six years, returning home during the summers to work with his family on the farm. His parents could not afford to support him fully during the school year; their

Figure 1.5. The Troppau (Opava) Gymnasium and Church of St. Adalbert, where Mendel attended boarding school, in the years 1834–1841, from the time he was twelve years old until age eighteen. The school was in the tall section of the building in the center of the image, attached to the church in the background. *Drawing by Daniel J. Fairbanks from a historic photograph, sepia ink and pastel on paper. Collection of the artist.*

payments covered only bed and half rations of food, forcing him to rely on sporadic deliveries of food carried by family members and other villagers who were traveling to Troppau. This situation lasted for the first four years. However, as he recalled later (writing in the third person), "due to several successive disasters, his parents were completely unable to meet the expenses necessary to continue his studies, and it therefore happened that the respectfully undersigned, then only sixteen years old, was in the sad position of having to provide for himself entirely."[7]

Committed to continuing his education, he enrolled in a course to certify as a *Schulkandidaten und Privatlehrer*, a private tutor for other students, as a means to financially support himself by receiving payment for one-on-one instruction. The tutoring payments provided, as he later put it, "a scanty livelihood."[8] However, the stress of this situation drove him to serious illness, and he retreated home to recover in 1839. Returning to school, he made up for lost time and finished his schooling, graduating near the top of his class at the age of eighteen in 1840 with a diploma known the *Abgangszeugniss*.[9]

In his late teens, while studying at the Troppau Gymnasium, Johann penned two related poems, handwritten in somewhat crude German but genuinely heartfelt (here in English translation):

To what end was man created?
To what end did an unfathomably great being
Breathe life into that bit of dust?
Certainly, the Most-High,
Who so knowledgeably shaped the earthly sphere,
Who not without purpose
Brought forth the worm in the dust,
Did not create man for nothing;
Certainly, the capabilities the mind received
Reveal that a lofty goal is its destiny. —
Ceaseless effort, refinement and cultivation of his power,
Such is the lot of man here below.
But the laurel for one who earnestly and ardently
Pursues the cultivation of the mind,
Who seeks and finds the secret depths of knowledge
With lightning flashes of understanding,
In whose developing power
The seed of most glorious invention plants itself,
Nourishes and finally sends its blessing
To the needy human throng in blessed bounty.
Yes, that laurel never withers,
Even when the whirlpools of time
Pull generations into the abyss,
Even when only mossy ruins
Remain of the time when the genius appeared. —
How gloriously will the shifting tides
Of jubilation's song resound!
How keenly will fame strive
To celebrate and crown this work!
What depth of gratitude will the voices of future generations
One day grant to the art of Guttenberg! —

You letters, offspring of my research,
You are the solid rock on which I will, for eternity,
Erect and fortify the temple of my fame.
You are meant, by the master's wish,
To dispel the dark power of superstition
That weighs heavy on the earth;

To bring to light and conserve
The works of the greatest men,
Works utilized by a scant few
Fallen prey to decay.
In many a head, slumbering and still hidden,
Your influence will develop the great, bright mind.
In short, your being should and will
Create a better, new life.
The greatest pleasure of earthly joy,
The highest destination of earthly bliss
Would be awarded me by destiny's power
If I, risen from my grave, were to see
My art flourishing in the midst of posterity![10]

These poems reveal his religious devotion, his dedication to education of the mind, and his admiration for the spread of knowledge through the "art of Guttenberg," an allusion to mass printing as the vehicle of knowledge. In several ways, these poems portend aspects of the life he would lead. By this time, he was considering becoming a priest, following in the footsteps of Pater Schaumann, his school's headmaster.

Things were not well at home. One of the "several successive disasters," as he put it, that had struck his family was a serious accident in 1838. While his father was working to meet his feudal labor commitment in lieu of monetary taxes, a large log rolled onto his chest, permanently injuring him. By 1840, the lingering disability was still present, magnifying the tension young Johann had faced throughout his early life. Should he relieve his father's burdens and take over the farm, or should he return to his education? Again, with his family's encouragement, he chose education.

To qualify for admission to the priesthood or to pursue university studies, he required two more years of education at a philosophical institute. The nearest one was in Olmütz (Olomouc), a sizable city that had been the capital of Moravia until its near destruction in 1640 during the Thirty Years' War. Johann enrolled there in 1840, and, again, he sought students to tutor as a means for supporting himself (figure 1.6).

Unfortunately, as he later described the dilemma, "all his efforts remained unsuccessful because of lack of friends and recommendations."[11] Moreover, many of the students were native Czech speakers, a language Johann was learning but in which he still lacked proficiency. While increasingly struggling to make ends meet, he attended classes through the semester of 1840–1841 and completed the final examinations in Latin and

Figure 1.6. The Philosophical Institute in Olmütz (Olomouc) that Mendel attended in 1841–1843. *Drawing by Daniel J. Fairbanks from historic photographs, sepia ink and pastel on paper. Collection of the artist.*

mathematics. However, an insidious nervous attack overcame him before he completed the remaining examinations, causing him to abandon them. By doing so, he forfeited the entire semester. Once again, he retreated home to recover.

This second bout of illness was devastating, leaving Johann with an uncertain future. If he returned to the Philosophical Institute, he would have to begin anew, as if he had never attended before. Nonetheless, he reaffirmed his commitment to continue his educational pursuits with the goal of becoming a priest. Anton, accepting his son's decision, sold the farm to Alois Sturm, husband of his daughter Veronika. The sale was in August 1841 while Johann was still at home recovering. The text of the sales contract provides a sense of where things stood with Johann, including evidence of his decision to pursue priesthood as his profession: "The purchaser shall pay to the son of the seller, Johann by name, if the latter as he now designs should enter the priesthood, or should he in any other way begin to earn an independent livelihood, the sum of 100 fl. . . . and shall also defray all expenses connected with the first mass."[12] The contract further included a small ongoing contribution to support Johann's continued education as well as a fallback provision that should he again return home, he would be given space to live and arable land to cultivate without charge.

Part of the contract for the sale of the home included a contribution toward Theresia's dowry. At the young age of eleven, she relinquished this income to her brother to help pay his educational expenses. On Johann's return to Olmütz in the fall of 1841, his financial situation had improved. He augmented the meager sustenance his family provided by successfully recruiting students to tutor. This time, he would succeed.

The topics he studied—and one he did not study—at the Philosophical Institute would dictate his future in ways he could not anticipate. In the sciences, mathematics, physics, and natural history were listed as obligatory. He excelled in both mathematics and physics, receiving the highest grades. However, the professor who taught natural history, Jan Hecelet, was suffering from a prolonged illness, requiring cancellation of the courses in this topic. Johann was apparently excused from this requirement, creating a deficit in his knowledge that would later haunt him. In 1850, when he undertook a grueling examination for teacher certification, he succeeded in physics but fell short in the natural history portion.

Johann's most influential teacher at the Olmütz Philosophical Institute was his physics professor, Friedrich Franz, who was highly impressed with the young man's academic performance. The textbook Franz used to supplement his lectures was by Andreas von Baumgartner, a professor at the University of Vienna who had previously held the chair in physics at the Olmütz Philosophical Institute. Baumgartner would later, in 1851, secure Mendel's admission to the University of Vienna. Franz, as a canon of the Premonstratensian Order, had lived at the St. Thomas Augustinian Monastery in Brünn, where he taught physics at the Brünn Philosophical Institute for nineteen years before moving to Olmütz in 1842. Johann was one of his first students in this new position. In an odd twist of serendipity, had Johann not fallen ill in 1841 and restarted his studies anew, Franz may not have encountered him. Now Franz would play the key role in guiding his future.[13]

Franz received notice from a member of the St. Thomas Monastery in the summer of 1843 requesting that he recommend candidates for admission. Johann was the only candidate he recommended, writing a glowing letter of support that has been preserved, its date July 14, 1843. Franz wrote that Johann "achieved top marks nearly continuously in both school years at the Philosophical Institute and is of very solid character . . . in my area of expertise he is probably the best."[14] The letter is addressed to "Honored colleague and very dear friend," without naming the recipient. Franz's letter asks that it be conveyed to the abbot, so, although he did not directly send it to the abbot, it probably made its way into his hands.

Of the thirteen applicants, Johann was one of four accepted as novices. The abbot, Cyrill Franz Napp, apparently was so impressed with Johann's qualifications and Franz's recommendation that he accepted Johann without the traditional interview, notifying him of his acceptance on September 7, 1843. To complete his admission, Johann required a letter of consent from his parents and approval from the archbishop of Olmütz, who oversaw Johann's religious participation at the time, and also a medical examination attesting to good health. This latter requirement could have been problematic given Johann's propensity to repeatedly fall ill under nervous stress. However, the physician in Odrau (Odry), a town to the north and within walking distance from Johann's hometown of Heinzendorf, examined Johann and gave him a clean bill of health.

On October 9, 1843, Johann was inducted as a novice into the Augustinian Order of St. Thomas and ceremoniously robed in the Augustinian habit. He took on the monastic name "Gregor" (the Germanized version of the Latin name Gregorius) and from this point forward used it as his first name. He was twenty-one years of age.

A Community of Scholars

2

a station of life which would free him from the bitter struggle for existence

—GREGOR MENDEL

THE BUILDINGS OF ST. THOMAS MONASTERY, where Mendel made his new home in the fall of 1843, had previously been a Cistercian nunnery attached to the Basilica of the Assumption of the Virgin Mary, a towering fourteenth-century Gothic church. The building complex resides at the foot of the Spielberg Hill on the western outskirts of Brünn, beyond the old city walls, in a place known as Altbrünn (Staré Brno), meaning Old Brünn (figure 2.1).

Figure 2.1. The St. Thomas Monastery in Altbrünn (Staré Brno) drawn to resemble its appearance when Mendel resided there. The buildings in the foreground are the monastery, a former Cistercian nunnery. The Basilica of the Assumption of the Virgin Mary is a Gothic church attached to the monastery, behind the monastery buildings in this image. The Spielberg (Špilberk) hill is in the background on the right, with the Spielberg (Špilberk) Castle on its top. *Drawing by Daniel J. Fairbanks from historic photographs, and sketches and photographs on site, sepia ink and pastel on paper. Collection of the artist.*

The Augustinian Order of St. Thomas had at one time, until the late eighteenth century, occupied sumptuous quarters in the central part of Brünn, adjacent the magnificent Church of St. Thomas. However, its members were forced to move in 1783 to the abandoned nunnery. This event and its aftermath make up quite a story, especially significant for Mendel in that it explains not only how he came to reside in the old monastery where he spent most of his life but also how a remarkable intellectual environment emerged within its walls.

During much of the eighteenth century, the Enlightenment transformed European and American history. Religious freedom and tolerance, basic human rights, education, and individual liberty were among its most relevant undercurrents. It motivated the American and French revolutions as royal dominion crumbled under the surge of democracy. When Austrian Empress Maria Theresa died in 1780, her son, Joseph II, ascended to the throne at the age of thirty-nine as Holy Roman emperor and archduke of Austria. His tumultuous rule would last merely a decade (1780–1790). The beginnings of the French Revolution were fomenting in Paris, and the American Revolutionary War was already under way. Joseph II was much enamored with Enlightenment philosophy and unilaterally mandated radical change throughout the empire under his rule. Marie Antoinette, his sister, met her final moment at the blade of a guillotine during the Enlightenment-motivated French Revolution.

The reign of Joseph II epitomized what is now known as "enlightened despotism," a monarchy headed by a ruler who retains absolutist control but enforces Enlightenment ideals by edict, often to consolidate power through a purported benevolent autocracy. Joseph II moved hastily, even recklessly, to transform his empire with reforms consisting of numerous edicts. He disbanded serfdom, granting legal rights to peasants, and abolished rent paid by servitude. He promoted education, requiring all children, female and male, to attend schools, asserting state control over schools and opening education to non-Catholic youth. In a land dominated by Roman Catholicism, he issued the Patent of Toleration of 1781, which granted increased freedoms to Lutherans, Calvinists, and Orthodox Serbs. The following year, he issued the Edict of Tolerance, which extended additional freedoms to Jews. As Holy Roman emperor, he asserted control over the Austrian Catholic hierarchy, attempting to weaken its ties to the pope and affirm its allegiance to the crown. These actions prompted a visitation by Pope Pius VI to Vienna in 1782.

Among the most influential reforms implemented by Joseph II was the closure of hundreds of monasteries, nearly a third of those existing at the beginning of his reign. He imposed secularization on the remaining monastic orders, requiring their members to serve the people as schoolteachers and to minister to the needs of the ill and the impoverished. He banned several orders, including the Capuchins, Carmelites, Dominicans, Franciscans, and Jesuits. He spared the Augustinians because of the order's mission of service, scholarship, and education, which were more in line with his ideals. Although many of his reforms met with catastrophic failure, their remnants lingered for decades to come. Their collective influence became known as Josephinism, which persisted to the time of Mendel's induction into the St. Thomas Monastery.

It was Joseph's decree that forced the St. Thomas Augustinians to move in 1783 from their sumptuous edifice in the city center to the abandoned nunnery in Altbrünn, consisting of dilapidated fourteenth-century buildings. Necessary renovations would leave the once-affluent monastery in a dire financial situation for decades to come. These buildings, magnificent yet in constant need of repair, would be Mendel's home for most of his life and the site of the garden and greenhouse where he would conduct his famous experiments.

The monastery was known as the *Königinkloster*, meaning "queen's cloister." The name honors Elisabeth Richeza of Poland, queen consort and dowager of Bohemia, who spent her final days in the Cistercian nunnery of Altbrünn, taking on herself the veil but not the vows. She financed the construction of the Basilica of the Assumption of the Virgin Mary during the years 1323–1334, the towering Gothic church attached to what was then the nunnery. She died in 1335, shortly after the church was completed. To this day, her remains lie beneath its floor. This church is where Mendel attended services and where his funeral would be held.

The name "cloister" was, in fact, a misnomer for the St. Thomas Monastery during the time Mendel lived there. Josephinist reforms had established a scholarly and educational climate where Augustinians were free to pursue secular scholarship and science. This enlightened intellectual climate thrived in the monastery, where its learned members attracted esteemed scholars from around Europe who often lodged in the guest quarters.

The vestiges of Josephinism continued to influence the Austrian Empire through much of the nineteenth century, especially at the University of Vienna. Catholic opposition to Josephinism fomented ongoing conflict between scientists and the church, a conflict Mendel experienced in Vienna as well as Brünn. He would repeatedly struggle with this conflict throughout his career.

In April 1850, Mendel wrote a brief account of his life up to that point in time as part of an application to be examined for teacher certification. Some authors refer to this account as "Mendel's autobiography."[1] However, it differs from a traditional autobiography in that Mendel wrote it when he was twenty-seven, so it accounts for only the early part of his life. He included selected events, occupying a mere four handwritten pages. Mendel called it *eine kurze Skizze*, meaning "a short sketch." One of its most quoted passages is Mendel's assertion that his financial situation was the principal circumstance compelling him to commit the rest of his life to the priesthood:

> The respectfully undersigned realized that it was impossible for him to endure such exertions any further. Therefore, after having finished his philosophical studies, he felt himself compelled to step into a station of life, which would free him from the bitter struggle for existence. His circumstances decided his vocational choice. He requested and received in the year 1843 admission to the Augustinian Monastery of St. Thomas in Altbrünn.[2]

In this case, the translator, Anne Iltis, employed the English word "compelled" for Mendel's original "*gezwungen.*" Klein and Klein have argued that a better translation of this word is "forced," which implies that Mendel might have chosen a different path had financial deprivation not made the priesthood his only tenable option.[3] Questions about his motivation for joining the Augustinian Order have abounded. Some have claimed that he did so reluctantly, motivated only by financial needs, others that he did so out of sincere religious devotion. There is evidence to support both arguments, which are not mutually exclusive. Mendel's true motivation probably included aspects of both.

He clearly understood that the priesthood would open a path for him to escape poverty. However, he could not have foreseen how extraordinary the educational and scientific opportunities were to be. His religious devotion is evident in his aforementioned poetry, fragments of sermons he gave later in life, and the fact that he remained dedicated to his vows, even when one of his closest friends in his order did not. Nonetheless, his deportment was far from the archetype of a secluded and pious monk. A private letter Mendel wrote reveals his disdain for the tedium of ceremonial rituals required of him, lamenting that he might as well "piss into the wind" were he to attempt to avoid them. He often dressed as other men when in public rather than consistently wearing his priestly vestments. And he spent his earnings to participate in expensive tours on what were advertised at the time as luxurious pleasure trains, one a tour of France, Germany, and England, another a tour of Italy, where he briefly met Pope Pius IX as part of a group audience.[4]

More serious were his acts of obstinate rebellion against what he perceived as unjustified demands and regulations. One of Mendel's colleagues referred to him as "learned and free-thinking."[5] He maintained a strong devotion to science and secular learning, which incensed his superiors beyond the monastery walls but harmonized with his fellow friars and his abbot. The available evidence indicates that Mendel was passionately dedicated to science yet also fully committed to his vows and the duties of his religious profession.

Biological evolution was already a popular topic for study and discussion among Mendel's associates due in large part to its promotion by professors and students at the nearby University of Vienna. Darwin, although a respected naturalist in England at the time, had yet to publish *Origin of Species* and was not widely known south of the English Channel. Biological evolution was mostly despised in religious circles but not at the St. Thomas Monastery. It was a popular theme, embraced with acceptance and discussion, both in the monastery and in the meetings of the local academic societies, which several of the friars, including Mendel, regularly attended.

This intellectual climate within the monastery, initially established through Josephinist reforms, had the strong support of Mendel's abbot, Cyrill Franz Napp. Born Franz Napp in 1792, he had been raised in a German-speaking home in Gewitsch (Jevíčko), west of Olmütz and north of Brünn in Moravia. His schooling was in Olmütz, including the same Philosophical Institute where Mendel studied years later. As a child, he spoke both German and Czech, becoming fluent in both, and studied Latin in school. As a priest, he gave stirring sermons in all three languages. He was accepted into the monastery in November 1810 at the age of eighteen, taking Cyrill (which he spelled with a double "l") as his monastic name in commemoration of the ninth-century Saint Cyril of the legendary saints Cyril and Methodius, who were the first to preach Catholicism in Moravia. As a member of the Augustinian Order, Napp immersed himself in academic studies, specializing in biblical criticism and ancient Near Eastern languages, including Arabic, Aramaic, Chaldean, and Syriac. He taught these disciplines at the Brünn Theological Institute beginning in 1816 and published on them.

After being elected abbot and prelate of the St. Thomas Monastery in 1824, Napp so enjoyed his teaching post that he refused to resign it, prompting the bishop of Brünn, Wencelas Urban von Stuffler, to forcibly remove him in 1827 from his teaching post as an incentive for him to devote greater efforts to his religious and administrative responsibilities. Napp continued to focus on teaching by accepting administrative appointments over Moravian schools. Throughout his abbacy, Napp cultivated a scholarly environment, seeking novices who had excelled in their prior education to serve as schoolteachers and professors in universities and institutes. He also encouraged professionalism in their various disciplines. When Mendel entered the monastery, Napp had been abbot for nineteen years, established as a seasoned and highly respected leader and scholar.

Philosophers in Mendel's day were typically experts in both theoretical and practical sciences, and Napp was no exception. He embraced the agricultural sciences, joining the Agricultural Society in Brünn and its sheep-breeding association and also the Pomological (fruit tree) Society, of which he would be elected president. He published the results of his scientific research in agriculture in the journals of these societies. He employed the latest advancements in science to improve the monastery's expansive agricultural holdings, which provided significant financial benefits to its coffers.

Under Napp's influence, the friars tended to be liberal in their political and scientific views, referring to themselves as freethinkers, often at odds with church authorities whose inclinations were more traditional. In addition to Napp, three of Mendel's fellow friars were advanced scholars and especially influential in their respective disciplines: Matouš František Klácel, Tomáš František Bratránek, and Pavel Karel Křížkovský (figure 2.2).

From Mendel's first days in the monastery, Klácel became his friend and mentor and over the coming years would have a profound influence on him. Like Mendel, he had entered the monastery under duress of poverty, reminiscing on his dire youth in later years:

> Where should I now go to? I, poor student? Mother died, who had found in my progress her only joy! Here I could not go, nor there, because of poverty. . . . I applied to the rich monastery of the Augustinian Friars in Brno, whose head [Napp] was a famous prelate, scientist, secret freethinker, and patriot, and expert in state affairs and economy. He accepted me kindly.[6]

Klácel was the most provocative of the friars. He was a philosopher, well educated in the natural sciences, and a follower of the philosophy of Georg Wilhelm Friedrich Hegel and the German *Naturphilosophie* movement. In later years, he would become a staunch proponent of Darwinism, especially in its application to human evolution.

Figure 2.2. Friars of the St. Thomas Monastery, probably 1862, nineteen years after Mendel joined the Augustinian Order of St. Thomas and during the time when he was conducting his famous pea experiments. Mendel, Bratránek, Klácel, Křížkovský, and Napp are labeled. *Period photograph, collection of the Mendelianum, Moravian Museum, Brno, Czech Republic, public domain.*

He was raised in a Czech-speaking home and later became an outspoken Czech nationalist. In 1827, he entered the monastery, chosen by Napp in the third year of his abbacy. In his first years at the monastery, Klácel studied philosophy at the Brünn Theological Institute and later at the University of Olmütz. Napp then assigned him to teach philosophy at the Brünn Philosophical Institute in 1835. Napp's immediate superior was the bishop of Brünn, Franz Anton Gindl, who served in that office from 1831 to 1841. Gindl avoided interfering with Napp's efforts to enhance the intellectual climate at the monastery and the schools.

However, when Gindl died in 1841 (two years before Mendel arrived), his successor, Anton Ernst Schaffgotsch, believed in strict orthodoxy. He was intent on rooting out remnants of the Enlightenment ideals of Josephinism, which he considered heretical. Schaffgotsch's wrath was kindled when he heard accusations that Klácel had been teaching secular and liberal philosophy. Despite Napp's protests, Schaffgotsch forced Klácel to resign his position in 1844 for teaching "pantheism and other heresies related to Hegelianism."[7]

Klácel spent time away from Brünn to escape the oppression, returning and leaving repeatedly over a period of years. After Mendel became abbot in 1868, Klácel deserted the monastery for good, possibly with Mendel's clandestine assistance, and emigrated in 1869 to the United States, where he remained in exile until his death. A magnificent monument to him stands at his grave in Belle Plaine, Iowa.[8]

Klácel's relationship with Mendel vacillated throughout his life. During their early years under Napp, the two were trusted friends. Later, years after Klácel had left the monastery, Mendel's nephew recalled an incident when he found a loose photograph of Klácel in a photo album. Mendel told his nephew to not affix the photograph in the album because "the priest had broken his vows."[9] Nonetheless, Mendel apparently thought enough of Klácel to keep the photograph. After emigrating to the United States, Klácel wrote the following account of Mendel's ascension to the abbacy, laced with mixed feelings of friendship and disdain: "Prelate Napp died at an advanced age—elected was a young professor, learned and freethinking, my friend, but a hypocrite, who knew how to please everybody . . . and swore onto anything the more willingly, the less he believed in it."[10] Notably, after Napp's passing, Mendel did not vote for himself as abbot but gave Klácel his vote in all three rounds.

Tomáš František Bratránek was also among the most scholarly and outspoken friars. Born in 1815, his mother was an ethnic German, his father Czech, and he was raised speaking Czech but later became a specialist in German literature and philosophy. He joined the St. Thomas Monastery in 1834 under Napp, who sent him to the University of Vienna, where he earned a doctorate in 1839. A philosopher, natural scientist, and follower of Hegelianism and the *Naturphilosophie* movement, Bratránek was called back to Brünn from Lviv in what is now Ukraine to assume Klácel's teaching post shortly after Mendel arrived. In 1851, he accepted a professorial post at the Jagiellonian University in Krakow, which was in Prussia at the time. Eventually, he would ascend to the prestigious position of dean of the faculty of arts and later university rector. He never severed his ties to the St. Thomas Monastery, returning there in retirement thirty years later, in 1881, when Mendel was abbot. He died there in 1884, a few months after Mendel's death. A proponent of evolution and a specialist on Goethe, Bratránek would compare Goethe's writings with Darwin's.[11]

An influential friar who joined the monastery shortly after Mendel (in 1845) and who became one of his most trusted friends was Pavel Karel Křížkovský, a choirmaster and composer of choral music. Mendel had known Křížkovský before either of them joined the monastery when they were students at the Troppau Gymnasium, then later at the Olmütz Philosophical Institute. Like Mendel, Křížkovský ended up abandoning his studies in Olmütz due to the stresses of poverty. Unlike Mendel, however,

Křížkovský did not return to the Olmütz Philosophical Institute, so he was required to complete his studies at the Brünn Philosophical Institute after entering the monastery. Although his surname was Czech, Křížkovský had been raised speaking German and before entering the monastery had spelled his name in German as Karl Krischkowsky.[12] He took on Pavel (Paul) as his monastic name and adopted the Czech spelling of his name for the rest of his life.

Discovering the beauty of Czech folk music, Křížkovský composed choral music based on Moravian-Czech folk songs. His success as a composer blossomed in the 1860s when he composed a cantata, *Sts. Cyril and Methodius*, in commemoration of the millennium of their arrival in Moravia. His choral song *Utonulá* ("The Drowned Maiden") received much praise when performed in Prague, where he received a major commission to compose an opera. However, news of such a prestigious yet secular commission incensed Bishop Schaffgotsch, who by then had investigated the St. Thomas Monastery for secularism. He prohibited Křížkovský from composing anything but church music, an order to which Křížkovský reluctantly acceded, abandoning his prestigious opera commission. A hymn he composed, *Ajhl'a, skvie sa oltár Pána* ("Behold, the Altar of the Lord Is Shining"), remains a lovely and favorite hymn in Czech Catholicism. He is especially known for compositions based on Czech folk songs, among them *Utonulá, Dar za laskú* ("Gift for Love"), *Odvendeného prosba* ("The Recruit's Prayer"), *Rozchodná* ("Song of Parting"), and *Dívčí* ("Maiden").

Křížkovský is perhaps best remembered as the first mentor of Czech composer Leoš Janáček, who enrolled in music studies under him in 1865 at the age of eleven while boarding at the monastery, the same year Mendel presented his famous lectures. In 1872, while Mendel was abbot, Křížkovský accepted the invitation from the archbishop of Olmütz to serve there as choirmaster, assigning Janáček, then eighteen years old, to assume his position as choirmaster at the Basilica of the Assumption of the Virgin Mary, attached to the monastery.[13] Křížkovský returned to the monastery the following year when things did not initially go well in Olmütz but shortly thereafter moved again to Olmütz, where he remained until 1883, when his health began to fail. He then returned to the St. Thomas Monastery, where Abbot Mendel was also in poor health. Křížkovský died in 1885, a little more than a year after Mendel's death.

In 1848, a wave of revolutionary sentiment enveloped much of Europe, especially the Austrian Empire. Abolition of the feudal system under which Mendel's family had lived was its aim, and the University of Vienna, the intellectual center of the empire, played a prominent role in the rebellion.

The St. Thomas Monastery was a hotbed where some of the friars vocally promoted revolutionary sentiments, including Abbot Napp. The author and poet Ludwig August Ritter von Frankl-Hochwart penned a rousing poem titled *Die Universität* ("The University") espousing the movement, which was printed as a flyer and widely disseminated. Křížkovský was one of several composers who set the poem to music.[14]

Six of the friars of the St. Thomas Monastery signed a petition demanding governmental reform to be delivered to the Brünn Provincial Diet, focused on granting greater rights and freedom to members of religious orders.[15] The petition contains scathing language, referring to the treatment of friars as "one step below a common criminal," that monasteries are "nothing more than seminaries for forced morals . . . for the poor and deluded young men," where "priests remain subjugated, contempt-exposed slaves, at whom the free and cultured nations of the world will be pointing with ridicule and disdain."[16] The final paragraph of the petition reads,

> Consequently, the undersigned professors and pastoral workers of the Order of St. Augustine in Altbrünn take the liberty of appealing to the imperial parliament to grant them constitutional civil rights, and request to be allowed to devote their entire efforts, according to their abilities and past services, to public teaching institutions and to free, united, and indivisible citizenship. The undersigned make it respectfully their missions to promote science and humanity in accordance with the spirit of constitutional progress.[17]

Mendel's signature appears as the fifth of the six at the end of the petition. Klácel's signature is first, so he is presumed to be its principal author, although the handwriting indicates that Mendel was the scribe.[18] Mendel's involvement was somewhat risky, for he had completed his probation just two days before signing. Although the petition reflected the liberal tendencies of his fellow friars, this action was among the first of many that would establish Mendel's reputation as an activist, even a rebel, cementing the approval and camaraderie of his immediate brethren while putting him at odds with more distant church and state authorities.

Mendel's novitiate period was one year, ending in September 1844. He then began his studies at the Brünn Theological Institute. Consistent with his past performance in school, his grades in all courses were the highest possible, as were his evaluations of diligence. Due to deaths and illnesses, by 1846, the monastery was in desperate need of new priests. Based on Mendel's exceptional scholastic performance, Napp determined to advance him to the priesthood early, before he had completed the prescribed years of study. The community of friars voted unanimously in support of this action. The day after Christmas in 1846, Mendel took his solemn vows, committing his life to "live in poverty and chastity until death according to the Rule of Our Holy Father Augustine."[19]

However, he could not be ordained until he had reached the age of twenty-five, which was still some months ahead. Napp wrote in a letter to Bishop Schaffgotsch that Mendel "has just completed the third year of his theological studies with praiseworthy success, and has always lived blamelessly, piously, and religiously."[20] After receiving all approvals, Mendel was ordained a subdeacon on the earliest day possible, his twenty-fifth birthday, July 22, 1847. The next steps came quickly. He was ordained a deacon on August 4 and a priest on August 6, with his first mass on August 15, 1847.[21]

As a priest, he continued his studies at the Theological Institute, which included religious history and law; ancient Greek, Hebrew, Chaldaic, Syriac, and Arabic; theology; biblical criticism; divinity; catechism; and pedagogy. He also studied agricultural sciences at the Brünn Philosophical Institute under Franz Diebl, whom Napp had recruited as a professor there and who coauthored several scientific articles on agriculture with Napp.[22] Mendel completed his studies and his probation in the summer of 1848, just two days before he transcribed and signed the revolutionary petition.

Mendel began his assignment as a parish priest that year, attending to parishioners who spoke both German and Czech, and he delivered sermons in both languages. Eventually, he was assigned to attend to the spiritual needs of patients confined to the St. Anna Hospital, just a few steps from the monastery. The nervous stress he had suffered as a youth reemerged with this assignment, and he again became seriously ill at the sight of suffering patients. Napp readily perceived the cause of Mendel's incapacitation. In 1849, a position for a teacher in mathematics and in Latin, Greek, and German literature became available in nearby Znaim (Znojmo), south of Brünn near what is now the Czech–Austrian border. Napp seized the opportunity to reassign Mendel. In his letter to Bishop Schaffgotsch explaining Mendel's new teaching assignment, Napp wrote that Mendel was "much less fitted for work as a parish priest, the reason being that he is seized with an unconquerable timidity when he has to visit a sick-bed or to see anyone ill and in pain."[23]

Mendel enthusiastically took on his new assignment, and by all accounts, he was an exemplary science teacher. However, in 1850, the Austrian government issued an edict that all teachers must be certified through a university-administered examination. The schoolmaster in Znaim recommended Mendel, along with two other teachers, for the examination at the University of Vienna that same year. The initial portion consisted of what was known as homework (*eingesandte Arbeiten*), two detailed essays that he was to complete at home over a period of eight weeks, one on meteorology and physics and the other on geology and the history of life. The examiner for Mendel's essay on meteorology and physics was Andreas von Baumgartner, a renowned professor of physics and chair of the commission for teacher examinations at the University of Vienna (figure 2.3). Baumgartner was author of the physics textbook Mendel studied during his education in Olmütz under Professor Franz, so he was already familiar with Baumgartner's approach. Baumgartner found Mendel's essay to be fully acceptable and well written.

Figure 2.3. Andreas von Baumgartner, professor of physics at the University of Vienna and Mendel's examiner in physics for his teacher certification examination in 1850, as he appeared in 1847, three years before Mendel entered the University of Vienna. *Drawing by Daniel J. Fairbanks, based on a lithograph by Josef Kriehuber, 1847, sepia pastel on paper. Collection of the artist.*

The examiner for the essay on geology and natural history was Rudolf Kner, a zoologist and paleontologist who had recently completed his first year as a professor at the University of Vienna. At age thirty-nine, Kner was still relatively early in his career (figure 2.4).

Figure 2.4. Rudolf Kner, professor of zoology at the University of Vienna and Mendel's examiner in geology and natural history for his teacher certification examination in 1850, as he appeared in 1852, when Mendel was a student in Vienna. *Drawing by Daniel J. Fairbanks, based on a lithograph by Josef Kriehuber, 1852, sepia pastel on paper. Collection of the artist.*

The text of Mendel's essay makes it clear that he had a decent grasp of geology and paleontology and that he understood basic evolutionary theory of his day. Nonetheless, Kner failed Mendel, claiming that he had neglected to cite up-to-date sources. Modern examination of Mendel's essay shows Kner's criticism to be mostly unjustified; the essay is well informed and accurate given the state of knowledge at the time. Two of

Mendel's biographers from the twentieth century, Hugo Iltis and Vítěslav Orel, speculated that Kner was probably biased against Mendel because of his membership in a religious order, even though Mendel's essay was entirely secular with an emphasis on biological evolution and no reference to supernatural creation.

In it, we find Mendel's first known allusions to evolution:

> In the course of time, when the earth had attained the capability necessary for the formation and preservation of organic life, plants and animals of the lowest species first appeared. The period of organic formation was not infrequently interrupted by catastrophes, which threatened the life of organisms and, in part, led to their decline. . . . The vegetable and animal life developed more and more richly; its oldest forms disappeared in part to make way for new and more perfect ones.[24]

The split decision did not disqualify Mendel; he could still attend the examination in Vienna, which consisted of an on-site written portion (*Klausurprüfung*) and an oral session (*viva voce*). Confusion arose, however, when a notice of rescheduling arrived too late at the monastery. The original examination date was at the beginning of university holidays in August 1850, and the rescheduled date was postponed until the next academic year in the fall. Unaware of the change, Mendel traveled to Vienna for the original date in August, just prior to vacations. Caught off guard as the professors were preparing to leave, he appealed for leniency and received permission to complete the examination albeit under less-than-favorable circumstances.

Baumgartner and the famous physicist Christian Andreas Doppler examined Mendel in physics, and both determined that his written essay met the standard they expected. Kner was again the examiner in natural history, and he questioned Mendel on the classification of mammals, a topic for which Kner was one of the world's leading experts. Mendel was ill prepared for this topic and wrote a seriously flawed essay on-site without an opportunity to consult sources. Kner failed him, though Mendel did not know this at the time. According to modern reviews of Mendel's essay, Kner's criticism this time was appropriate. Kner, nonetheless, recommended that Mendel be advanced to the oral *viva voce* examination.

Unfortunately, Mendel's *viva voce* performance was mediocre. According to his examiners in physics, "Though he has studied diligently, he lacks insight, and his knowledge is without the requisite clarity, so that the examiners find it necessary, for the time being, to declare him unqualified to teach physics in the lower schools."[25] Kner was more positive about

Mendel's performance: "the candidate showed more knowledge and gave evidence of more diligent study than might have been expected from the written papers." Kner nonetheless concluded "that he is not yet competent to become a teacher."[26]

Announcement of the result was delayed because of the holidays, possibly leaving Mendel wondering. The lack of certainty was fortuitous in that it allowed him to serve as a substitute teacher at the Brünn Technical Institute, beginning in April 1851. Although the written result of the examination was completed on October 17, 1850, a series of delays prolonged its arrival. The report was first sent to the Moravian Regional Educational Authority, which, in turn, sent it to the directorate of the Brünn Gymnasium. The report finally reached Mendel on August 9, 1851. By then, Napp had already contacted Baumgartner to arrange for Mendel to begin studies at the University of Vienna.

Mendel undoubtedly suffered dejection on receipt of the results. Although he did not realize it at the time, his partial failure was serendipitous, the impetus that would make possible an extraordinary university education in Vienna, one of the world's great centers for the sciences. According to Vítězslav Orel,

> Had he passed his teachers' examination, he would have stayed on at the Znojmo [Znaim] Gymnasium, and several generations of secondary school pupils would have gained an excellent teacher. Science, on the other hand, would almost certainly have lost one of its leading discoverers.[27]

The commission recommended that Mendel reapply for the examination after at least one year had passed. Abbot Napp, in consultation with Andreas von Baumgartner, arranged for Mendel to spend a year at the University of Vienna to prepare him for this examination. Mendel's actual time in Vienna extended into almost two years, encompassing the academic years of 1851–1852 and 1852–1853, providing him with an even richer university education than originally anticipated.

Baumgartner wrote to Napp requesting that the monastery provide full financial support for Mendel's course of study, a recommendation that Napp enthusiastically endorsed. In an attempt to justify the expense, Napp wrote to Bishop Schaffgotsch that Mendel had "exceptional intellectual capacity and remarkable industry in the study of natural sciences, and his praiseworthy knowledge in this field has been recognized by Count Baumgartner."[28]

The bishop agreed but with the caveat, consistent with Schaffgotsch's mistrust of the St. Thomas Augustinians, "that in Vienna the above-named priest shall lead the life proper to a member of a religious order, and shall not become estranged from his profession."[29] This sentiment would come back to haunt Mendel five years hence, when Schaffgotsch would accuse him of the very actions presaged in this warning.

The University of Vienna was then (and is now) one of the premier universities in the world. Mendel had the good fortune of working under the tutelage of scientists who were leaders in their specialties, particularly in mathematics, physics, and botany. While in Vienna, he would earn the respect of his professors, as evidenced by his election to the prestigious Imperial-Royal Zoological-Botanical Society in Vienna, where he presented and published his first scientific research. This extraordinary university education would establish the course of his scientific career and set the stage for his groundbreaking discoveries that, at the opening of the twentieth century, would posthumously bring him worldwide fame.

The Evolution of a Scientist
Mendel in Vienna

3

So attached to the study of nature

—GREGOR MENDEL

THE UNIVERSITY OF VIENNA in Mendel's day was one of the largest and most prestigious universities in the world, and the timing of his enrollment could hardly have been more fortuitous. The 1848 revolution brought about major reforms in the university with an increased focus on science, especially the experimental foundation of scientific research and discovery. These reforms supported the growing industrial and technical expansion of the city and the empire. The university was in the process of expanding from its central location at the *Universitätplatz* to buildings on the city's periphery where scientific institutes were established (figure 3.1).

The university recruited some of the most renowned scientists in all of Europe to establish these scientific institutes where students studied in small classes with hands-on experimentation. Mendel studied under such luminaries as Christian Doppler, the physicist and astronomer after whom the Doppler effect is named; Andreas von Ettingshausen, who pioneered the design of electric generators and developed theoretical combinatorial mathematics; and Franz Unger, a botanist and paleontologist who advanced cell theory in biology and whose pre-Darwinian inferences and teachings on evolution were remarkably aligned with those Darwin would publish several years later.

As exhilarating as it must have been for Mendel to begin his studies at the University of Vienna, his first days there were tumultuous. The necessary arrangements for him to depart the monastery and to fund his educational and living expenses had consumed far too much time, causing him to arrive well beyond the enrollment date. By the time things were settled and he finally departed Brünn for Vienna, he was five weeks late for the start of the fall semester.

Bishop Schaffgotsch had insisted that Mendel lodge at a monastery to "lead the life proper to a member of a religious order."[1] According to a letter written by Napp, after a failed attempt to arrange accommodations

Figure 3.1. The Aula der Alten Universität (Assembly Hall of the Old University), the largest university building overlooking the Universitätplatz, an area near the center of Vienna where several buildings of the old university were located. Mendel registered and attended some lectures here. While Mendel was attending the University of Vienna, it was expanding from its central location in the city into buildings on the city's periphery, where he also attended classes. The Universitätplatz is mostly unchanged today from its appearance when Mendel was there. This building currently houses the Austrian Academy of Sciences. *Drawing by Daniel J. Fairbanks, from sketches and photographs on site, sepia pastel on paper. Collection of the Mendelianum, Moravian Museum, Brno.*

for Mendel in the Monastery of the Brothers of Mercy, "My only recourse was to send Pater Gregor to Vienna with instructions to fend for himself, and to seek board and lodging in some other monastery or religious house. He left for the capital by the night train on October 27th."[2] Mendel initially stayed in an inn but shortly thereafter rented long-term living quarters in a building divided into apartments.[3] It was a commercial enterprise operated by the nuns of the Order of St. Elizabeth to supplement their coffers. Despite its ownership, it was not the type of monasterial accommodation that Bishop Schaffgotsch had in mind. Instead, people of many walks of life resided within its walls.

Immediately on arrival, Mendel contacted Baumgartner, who assisted him with late admission. Mendel's late arrival was, in fact, serendipitous. The Physics Institute had recently moved into another building, and the move had delayed the start of classes until almost the day when he arrived. Only one course was available to him on the delayed schedule that first semester, and it would be pivotal. The renowned physicist Christian Doppler, who arrived at the University of Vienna only a year earlier, had been named as the inaugural director of the university's newly established physics institute, and he would be Mendel's first teacher (figure 3.2). The course was already enrolled to its capacity of twelve students, but Mendel was granted a special exemption to enroll as a thirteenth, in part because of his proven interest and expertise in physics.

Figure 3.2. Christian Doppler, professor of physics at the University of Vienna and Mendel's first teacher at the university, as he appeared at the time Mendel studied with him. *Drawing by Daniel J. Fairbanks based on a period photograph, sepia pastel on paper. Collection of the artist.*

Doppler's teaching approach focused on hands-on experimentation, and his students were required to devise experiments to test hypotheses. Unfortunately, no record remains of what experiments in physics Mendel conducted. Although some students published the results of their experiments, Mendel did not. He again enrolled in Doppler's course in the spring–summer semester of 1852. The course title was the same: "Demonstrative Experimental Physics." Illness, however, overwhelmed Doppler, and he was forced to leave Vienna after this semester, in 1852, retreating to a milder climate in Italy to convalesce. He unexpectedly passed away in March 1853 in Venice at the age of forty-nine, just under a year and a half from the time Mendel began studying with him. Mendel must have felt sorrow for his mentor, though no writings remain to inform us of his feelings at the loss.

Physicist and mathematician Andreas von Ettingshausen took Doppler's place (figure 3.3). His fame matched that of Doppler, and he also was a fine teacher. He had been at the University of Vienna since 1821 and assumed the institute's directorship at Doppler's passing. At the time, Ettingshausen was especially known for his contributions to combinatorial mathematics, having authored a widely read book on the subject. The fact that Mendel studied under one of the world's foremost leaders on this topic would be influential, for this was the type of mathematics he would later employ when designing and analyzing his famous experiments. He also took courses in mathematics: one on logarithms and trigonometry from Franz Moth, the other on higher mathematics from Ettingshausen.

Figure 3.3. Andreas von Ettingshausen, professor of physics at the University of Vienna, who taught Mendel physics and combinatorial mathematics, as he appeared at the time Mendel studied with him. *Drawing by Daniel J. Fairbanks based on a period photograph, sepia pastel on paper. Collection of the artist.*

Some have argued that Mendel's experimental contributions were unique because biology of the mid-nineteenth century was observational rather than experimental, which is an overstatement. Although observation of organisms in nature was standard practice of the time in biology (as it is today), there was plenty of biological experimentation dating to more than a century before Mendel's time in Vienna. Experimentation was especially prevalent in plant hybridization, the emphasis Mendel would pursue. His botany professors, Franz Unger and Eduard Fenzl, were experienced researchers, having conducted numerous botanical experiments. At the encouragement of Unger, Mendel carefully read the work of plant hybridists and their experiments, citing them numerous times throughout his famous article. Darwin likewise cited abundant hybridization experimentation in his books, including some of the same experiments Mendel cited. Darwin was also an experimentalist, including the results of his own experiments in his books and articles and relating them to the experiments of others.

Unger and Fenzl were widely known; both held prestigious chairs in honor of their accomplishments. Mendel seems to have been especially fond of Unger (figure 3.4), years later arguing in favor of Unger's observations, discoveries, and inferences of how plants reproduced and against Fenzl's views.

Figure 3.4. Franz Unger, professor of botany and paleontology at the University of Vienna, as he appeared in 1864. *Drawing by Daniel J. Fairbanks based on a lithograph by Eduard Kaiser, 1864, sepia pastel on paper. Collection of the artist.*

Unger was convinced that cells were the basis of organismal growth and that cells led to an unending chain of life from one cell generation to the next and from one sexual generation to the next. As a botanist and paleontologist, Unger attempted to link modern plants with those in the fossil record to portray an unbroken hereditary and evolutionary tree, branching from the ancient past to the present.

Unger did not confine his evolutionary teachings to his classes but was especially dedicated to promoting evolution to the general public. He published a series of articles on botanical evolution in the *Wiener Zeitung* (*Vienna Times*), Vienna's principal newspaper. He also authored books intended for laypeople in an attempt to popularize evolution over epic geological periods. Such action kindled the wrath of some members of the Viennese clergy who perceived Unger as an enemy of Catholic orthodoxy.

Unger taught that the earth and the history of life on it were very old, immensely older than a literal interpretation of the Bible suggested. Contrary, however, to the prevailing view of "centers of creation," espoused by Darwin's friend and mentor, Charles Lyell, and the famed French geologist George Cuvier, Unger taught that species were not fixed and that present species had arisen from older ones. Science historian Sander Gliboff named Unger's theory of evolution the "theory of universal common descent" because it proposed that all current species are linked with previous ones through hereditary descent.[4] Unger's theory was remarkably similar to the one Darwin would propose eight years later in *Origin of Species*. Unger and Darwin viewed the emergence of new species not simply as an unbroken line connecting each species to its predecessors but as a branching tree diversifying into multiple species from single common ancestral species. As Unger put it, evolution was "not by any means a one-sided lineal progression, but a radiation broadening out on all sides."[5]

Mendel owned a textbook coauthored by Stephan Endlicher and Unger (1843) titled *Essentials of Botany* (*Grundzüge der Botanik*), and Unger authored the third part of the book, which Mendel extensively annotated. Given its publication date of 1843, it is unknown whether Mendel acquired the book before or after studying under Unger. In this book, Unger attributes the "*vis vitalis*" (vital force) of plants to chemical substances, an assertion Mendel would reinforce in his classic 1865 lectures.[6]

From May to October 1851, shortly before Mendel arrived in Vienna, Unger had published his series of articles as supplements in the *Wiener Zeitung*. He collectively named them *Botanical Letters* (*Botanische Briefe*) and left the authorship anonymous. The articles were shortly thereafter compiled and published in book form with Unger's name as author[7] and soon translated and published in English.[8] Unger's intent with *Botanical Letters* was to popularize evolutionary theory and his focus on plants but ultimately to extend the theory to humans in the final installment. In the first of these letters, Unger foreshadowed the theme of Mendel's experiments while arguing against the fixity of species: "who can deny that new combinations of the elements arise out of this permutation of vegetation, ever reducible to a certain law—combinations which emancipate themselves from the preceding characteristic of the species and appear as new species?"[9]

Unger's articles incensed Sebastian Brunner, a renowned, vocal, and controversial scholar and Catholic priest. Although his surname ancestrally ties him to the city of Brünn, he was born in Vienna and lived there his entire life. He was a gifted writer who had studied at the University of Vienna and for a time was professor of philosophy there. He authored voluminous books and articles, their subjects as wide ranging as biography, travel, religious history, anti-Semitism, satire, and poetry. His fundamentalist orthodoxy led him to confront the teachings of professors at the University of Vienna. He took it on himself to vigorously safeguard Viennese Catholicism against secularism, liberalism, Judaism, and Enlightenment ideals, with a specific disdain for the remnants of Josephinism. Brunner's friends and enemies alike called him by the Latin moniker *Malleus episcoporum*—the "Bishop's Hammer."[10]

Reacting against the 1848 revolution supported by Mendel, Brunner established the *Vienna Church-Times* (*Wiener Kirchenzeitung*), a newspaper focused on Catholic news and topics, serving as its principal editor from its founding in 1848 until 1865. When Mendel arrived in Vienna in the late fall of 1851, Brunner was a rising figure in the city, his *Wiener Kirchenzeitung* barely three years old. Unger had just published the final installment of the *Botanical Letters* series in the *Wiener Zeitung*, noting that the evolutionary process told the history not only of plant and animal life but of humankind as well. As Unger put it in the last sentence of his final article, "The slumbering world spirit [contextually meaning the unbroken lines of inheritance over eons of time] . . . in man sings its final hallelujah."[11] One week later, October 25, 1851, Brunner unleashed his first attack on Unger. He titled the article "Our Universities" (*Unsere Hochschulen*) and began with a condemnation of the secular trends at the University of Vienna, stating that "paganism is taught in the universities in all branches of science."[12] Turning his attention to Unger, Brunner co-opted this last line of his article: "Hence, we ask, amazed: who or what sings in man a hallelujah? . . . not the world, or nature-spirit, but that spirit which . . . constitutes a human."[13] Mendel arrived in Vienna two days later.

Unger had, before moving to Vienna, published a two-part book titled *The Primitive World in Its Different Periods of Formation* (*Die Urwelt in ihren verschiedenen Bildungsperioden*), visually depicting the evolution of life over long geological periods and culminating in the emergence of humans. Unger now prepared a new version in Vienna, published in 1851, the same year Mendel arrived there.[14] It was intended to popularize the earth's evolutionary history for the public. One part, printed in landscape folio format, consisted of large hand-tinted color lithographs of how plant and animal life appeared in each epoch as informed by fossils. The artist Joseph Kuwasseg made paintings illustrating the geologic epochs, and lithographer Leopold Rottman re-created them as lithographs for the book. The other part was an accompanying book with text in French and German to explain the images. The work was a grand achievement, a masterpiece of art and science, intended for tabletops in the living rooms of Europe's finest homes.

In the book's dramatic finale, Unger promotes a pantheistic evolutionary origin for humans as the apex of life evolving from simple to complex and from inferior to superior:

> Over much time, the creative forces had been employed for the production of numerous forms of animals and plants, always advancing from simple to complex, from roughed out masses to more noble beings capable of showing various expressions. Thousands of attempts emerged from their bosom, sketches always inferior to the desired creation. Finally, the great surge succeeded and man appeared, the work of a powerful and able master who wished to vivify the thought of the universe. Thus, we see him appear amidst the most varied beings, and of him alone we can say for the first time: "the word was made flesh."
>
> Truly there was no need to sow the marvelous seed of pure dragon teeth to call him into existence, for the germ of his life existed from the beginning to awaken only at the time destined for him! Delighted he looks upon himself, arising from the slumber of nature, and understands the purpose of his existence.[15]

Images from the book were displayed in Vienna in 1852 in popular "magic lantern" projection shows. The color images had been transferred to transparent glass plates mounted as slides in wooden frames. The magic lantern consisted of a set of lenses in a box, illuminated by limelight, which projected and focused the images on a screen in a theater. The show enhanced the fame of the paintings and widely promoted Unger's book. Mendel was in Vienna in 1852 when these shows began (figure 3.5). The images appeared in magic lantern shows held in London in 1861 after Darwin's *Origin of Species* had achieved widespread popularity there.[16]

Figure 3.5. A glass "magic lantern" slide of an image depicting plants during the Triassic period (approximately 220 million years ago), long before plants with flowers evolved. These slides were projected in vivid color onto large screens in theaters by bright limelight at "magic lantern" shows in Vienna, elsewhere in Europe, and in London to promote Franz Unger's book *The Primitive World in Its Different Periods of Formation. Photograph courtesy of Dr. Mark Wilson, Lewis N. and Marian Senter Nixon Professor of Natural Sciences, College of Wooster.*

Brunner lashed back at Unger in April 1852 in the *Wiener Kirchenzeitung*, again condemning the university and specifically branding Unger as "a man who openly denied the creation and the Creator."[17] Brunner's public attacks on Unger were, by now, so widely known that Mendel undoubtedly knew of them, especially as a Catholic priest. He nonetheless chose to enroll in two of Unger's classes in the fall of 1852, about a year after Brunner's initial attacks.

Although no written record has surfaced of how Mendel viewed Unger during the time he was in Vienna, there is ample evidence of Unger's impact on him. The strongest comes from Mendel's classic 1866 article wherein he strongly emphasized the fact that his discoveries supported "the view of famous physiologists" that "one germ cell and one pollen cell unite into a single cell that is able to develop into an independent organism."[18] Unger was undoubtedly one of these "famous physiologists."

Moreover, Mendel was a founding member of the Natural Science Society in Brünn, which listed Unger as one of its honorary members at the time of its founding.

As mentioned in this book's prologue, Unger was embroiled in a dispute with Fenzl at the time of Mendel's enrollment. Unger supported the view that female and male parents contribute hereditary elements equally to their offspring. Fenzl claimed that the pollen contributed all the inherited characteristics; nothing came from the female, and thus all inheritance was purely paternal. Mendel would later clarify in his article that his discoveries supported Unger's view, albeit without naming Unger, and he argued strongly against Fenzl's view, also without mentioning his name. According to Mendel's biographers Jan and Norman Klein, "Mendel's omission is puzzling. Was he, perhaps, afraid of acknowledging openly that he was in the same league as the pantheist Unger?"[19] Notably, Mendel also did not mention Darwin in his classic article, even though Darwin's influence is likewise evident.[20]

The idea that Mendel may have been reluctant to mention evolutionists in print is not so far-fetched. He presented his research to the Natural Science Society in Brünn not long after Bishop Schaffgotsch had carried out a prolonged investigation of the friars of the St. Thomas Monastery. The investigation concluded with Schaffgotsch's official censure and a recommendation that its members be disbanded for its rebellious secularism. In his censure, Schaffgotsch referred specifically to Mendel: "he studies profane sciences at a worldly institution in Vienna at the expense of the monastery to become a professor of said sciences at a state institution."[21] This latter comment reveals Schaffgotsch's opinion of the University of Vienna, a perception consistent with Brunner's attacks. Any mention of Darwin or Unger in Mendel's published writings could have further aggravated what was already a harrowing scenario.

Mendel's studies at the University of Vienna covered a broad range of topics. He enrolled as an "extraordinary student," meaning that he was not an ordinary degree-seeking student but rather one who audited his classes and did not receive formal performance evaluations. Therefore, his enrollment records list minimal information, such as the courses, the professors, and the number of hours per week for each course. He enrolled in two physics courses taught by Doppler and two by Ettingshausen, two botany courses with Unger and two with Fenzl, three courses in chemistry from Joseph Redtenbacher, one course in mathematics from Ettingshausen and another from Franz Moth, and two in paleontology from Friedrich Zekeli. He also enrolled in three zoology courses taught by Rudolf Kner, the same professor who had failed him in his certification examination. He became close friends with Professor Vincenz Kollar, who oversaw the insect collections at the university's Court Natural History Museum. Although Mendel did not take any formal courses from Kollar, he studied with him at the museum, and Kollar assisted him with his first two publications.

Few documents remain of Mendel's time in Vienna other than the two scientific articles he authored and published, his course records, and a few letters he wrote to members of his family and to fellow friars. His letters state little about his university studies, focusing instead on the weather, news of the time (such as a failed assassination attempt on the emperor), and lighthearted matters. One letter to a fellow friar, Anselm Rambousek, who joined the monastery as a novice at the same time as Mendel, reveals a somewhat unpriestly sense of humor that Mendel felt comfortable sharing privately with a friend. In the letter, he complained of having to return to Brünn to dress in regalia to attend a religious ceremony and that he needed new shirts for it because

> of the dozen shirts I brought with me to Vienna as many as 12 are frayed and in holes. . . . Would it not be a scandal if the new man I shall become in consequence of the pious exercises were to go about in a frayed shirt? How ashamed I should be if I (Apocalypse: *Stantes amicti stolis albis* [they stood clothed in white raiment]) had to parade in torn vesture![22]

Resigned to the fact he was required to leave Vienna for the ceremony, he sarcastically lamented, "it would be stupid of me to try to piss into the wind."

Scientists in nineteenth-century Europe typically were nominated and elected as members of local scientific societies based on their accomplishments. Each society served as a venue for discussing and disseminating research conducted by the society's members. The members typically assembled monthly, and each meeting featured a formal presentation of new research results, usually given by a member of the society. The proceedings of these meetings were published in a printed journal bearing the name of the society, including articles written by the presenters. Libraries subscribed to these society journals to make scientific research from around the world available to their patrons. Darwin, for example, first published his theory of natural selection as an article titled "On the Tendency of Species to Form Varieties" in the *Journal of the Proceedings of the Linnean Society* in 1858.

In 1851, shortly before Mendel arrived at the University of Vienna, several of his professors established a new scientific society: the Imperial-Royal

Zoological-Botanical Society in Vienna. Mendel was elected to its membership in January 1853, the final year of his studies in Vienna. This was a significant honor conferred by his professors who were founding members of the society—Unger, Fenzl, Kner, and Kollar—evidence of their esteem for him. Mendel's first two scientific publications appeared in the journal of this society and were reported in the *Wiener Zeitung*.[23] The first of these is the text of Mendel's invited lecture to the society: "On the Depradation of Radishes by Caterpillars" (*Über Verwüstung am Gartenrettig durch Raupen*) based on research he had conducted documenting the first known case of predation on radishes by larvae of the mother-of-pearl moth, known to consume other garden plants. This was his first scientific article.[24]

In the summer of 1853, having completed his university studies, Mendel moved back to the St. Thomas Monastery in Brünn. He turned his research attention to the pea plants growing in the monastery garden that summer, sending another manuscript to the same society for presentation and publication. Mendel was unable to attend the meeting where his paper was to be read. Instead, Kollar read the paper to the society on Mendel's behalf. The text of the presentation was published in 1854 as an article in the journal titled, "Description of the Pea Weevil, *Bruchus pisi*" (*Beschreibung des sogenannten Erbsenkäfers, Bruchus pisi*).[25] In this research, Mendel documented how he could identify seeds infested by the tiny pea weevil, newly emerged from the egg, by a minuscule pinhole in the surface of the seed coat, visible only under magnification. Once inside the seed, the weevil continued to consume the seed from its tiny lair inside of it, ultimately growing to a larger size while creating an increasingly larger hole hidden within the seed. Inspectors examined pea seeds intended for market, and if they found weevil infestations, they prohibited the sale of such peas, resulting in economic losses.

Interestingly, a decade later, infestations by this same weevil would make it impossible for Mendel to continue his famous experiments, prompting him to end them and prepare a summary of his research for presentation and publication. These two articles, published in a respected Viennese journal, were the first of seven journal articles Mendel would publish between 1853 and 1870, his classic article as the third chronologically.

Mendel's studies at the University of Vienna ended in July 1853. He returned to Brünn on July 21 to resume his duties at the monastery. He had spent almost two years studying in Vienna with some of the most respected scientists in all of Europe, attaining a broad university education in the sciences and mathematics. He had done so with the express purpose of preparing for certification as a teacher. It therefore is puzzling that he did not promptly schedule another certification examination. Instead, he returned to Brünn to immediately suffer censure from Bishop Schaffgotsch for pursuing a university education. The coming few years were to be especially troubling but also among the most significant as Mendel planned and initiated his epic experiments.

The Methodical Origins of a Masterpiece

4

blue eyes twinkling in the friendliest fashion through his gold-rimmed glasses

—FORMER STUDENT REMINISCING
ABOUT HIS TEACHER, GREGOR MENDEL

MENDEL'S INTELLECTUAL BLISS at the University of Vienna abruptly ended on his return to Brünn on July 21, 1853, when he arrived to find his monastic community in turmoil. Pope Pius IX had issued a decree that Austrian monasteries be purged of secular philosophy and science, authorizing Prince Friedrich Johann Jacob Celestin von Schwarzenberg, cardinal of the Holy Roman Church and archbishop of Prague, to carry out "visitations" at monasteries. The cardinal assigned the task for the Augustinian monasteries in Prague, Krakow, and Brünn to Bishop Schaffgotsch.

Abbot Napp was belligerent. By virtue of his status as prelate, he considered himself to be equal in authority to Bishop Schaffgotsch. He dispatched a strongly worded protest to Cardinal Schwarzenberg but to no avail. Bishop Schaffgotsch had already accused the friars of the St. Thomas Monastery of rampant secularism and pantheism. Now he had an official mandate, which he viewed as justification for taking extreme measures.

He enacted a prolonged investigation that dragged on for more than a year, beginning in February 1853 while Mendel was still at the University of Vienna.[1] The bishop delivered a written questionnaire to each of the members, asking probing, personal, and even offensive questions on the most private matters, such as a demand for written confessions on when and how many times each had broken his vows of chastity. On this point, not surprisingly, none of the friars made any confession. Schaffgotsch then conducted his personal visitation at the monastery on June 7 and 8, 1854, just shy of a year after Mendel had returned.

The bishop was thorough, leaving no stone unturned. He prepared a lengthy report, written in Latin, that he submitted to Archbishop Schwarzenberg in Prague on September 7, 1854. It could hardly have been more caustic. In it, he claimed that the "last ray of spiritual life" had departed from the monastery, a consequence of its members' focus on science and teaching, and that "in the house tending to the Rule of St. Augustine reigns a secular spirit which the few lappets of the Augustinian habit fail to cover up." Specifically, he reported that Abbot Napp held so many positions in societies and business entities that he was negligent of his abbatial duties, Klácel had "pantheistic fantasies," Neděle explained the Bible "in terms of the errors of the rationalists," Thaler "uttered blasphemic words in an intoxicated state in front of his students," Rambousek bathed "almost naked" in public, and Mendel studied "profane sciences at a worldly institution in Vienna at the expense of the monastery to become a professor of said sciences at a state institution."[2]

The bishop's choice to levy this accusation toward Mendel is revealing. Although he had supported Mendel's studies at the University of Vienna in 1851 for the purpose of teacher training, the bishop now viewed this experience as condemnable. Mendel had been officially appointed as a teacher at the *Realschule* on May 26, 1854, preceding Bishop Schaffgotsch's personal visitation by just twelve days. The bishop's accusation against Mendel makes it obvious that this appointment was on his mind during the visitation. Despite this criticism, Mendel would remain in this teaching position for almost fourteen years, leaving it only when he was elected as abbot in 1868.

Some of Bishop Schaffgotsch's accusations were seriously outdated, particularly those accusing Philipp Neděle and Aurelis Thaler, both of whom had died more than a decade before the report. According to Mendel's biographer, Vítězslav Orel,

> The bishop complained that not a single member of the community was willing to admit the errors of his ways, and that they stood behind their misguided superior [Napp] to a man. They even had the nerve to ask for a change in the rules of the monastery which would allow them to devote even more time to science and teaching.[3]

Schaffgotsch insisted that the monastery was beyond hope and must be dissolved. The members, under Napp's direction, humbly appealed to Cardinal Schwarzenberg in Prague, Mendel among them, detailing the changes they would make to bring their behavior in line. Their appeal was fruitless. Schwarzenberg endorsed Schaffgotsch's report and, on December 27, 1855, sent his determination to the Vatican. He proposed that Napp be retired with a pension and that the remaining members be disbanded and reassigned to other monasteries.

At this point, the whole ordeal recedes into a mysterious fog. No record remains at the Vatican or at the monastery of any action taken in any direction—no evidence of closure, reprimand, edict, exoneration, or even a letter of acknowledgment—apparently nothing but silence.[4] In the meantime, the Augustinians of St. Thomas adhered more strictly to the rules of the order and the changes they had proposed, living under a dismal cloud. They knew dissolution of their community could come at any time.

Given his love of teaching, Mendel was pleased with the offer from the *Realschule*. The school had been established while Mendel was in Vienna in 1851 as a post–elementary school, the equivalent of a *Gymnasium* at the middle and high school levels for students in their teenage years. Its focus was on the sciences to prepare students for technical professions, which were dramatically rising because of Brünn's economic growth as an industrial center. The school's director was Joseph Auspitz, who was already familiar with Mendel's teaching; he had been a professor at the Technical Institute and had recruited Mendel to teach there briefly in the spring of 1851 after Mendel had completed his appointment in Znaim and before enrolling at the University of Vienna. Auspitz penned a powerful letter arguing that, although Mendel had yet to be certified, he had studied physics and natural sciences at the University of Vienna and that Mendel's teaching at the Technical Institute had been exceptional. The governing body, known as the *Landesschulrat*, accepted Austpitz's arguments, allowing Mendel to assume the teaching appointment albeit in the role of a supply (substitute) teacher at half salary.

Mendel, for unknown reasons, had not scheduled another certification examination in 1853 when he finished his studies in Vienna. With the invitation to teach, the lack of certification returned to haunt him. According to the monastery's financial records, he traveled to Vienna in April 1855 to schedule the examination. As before, it would be in three parts: a homework portion (*eingesandte Arbeiten*), an on-site written examination (*Klausurprüfung*), and then an oral session (*viva voce*). Mendel completed the homework phase sometime between April 1855 and May 1856, although the date is not known. Unlike his essays from 1850, which were preserved in the university's records, his essays for this second examination are missing. According to Hugo Iltis, one of Mendel's earliest biographers, a note in the University of Vienna's archives "informs us that the relevant papers were destroyed."[5] Evidence that Mendel successfully completed the homework portion is in a letter written in Czech on May 8, 1856, by Klácel addressed to Bratránek stating that "his homework, etc. was graded as excellent."[6] Having completed the first stage, Mendel scheduled the on-site written and oral sessions. He traveled to Vienna and sat for the on-site written examination on May 5, 1856.

In an interview conducted more than sixty years after the fact, Johann Nowotný,[7] who was one of Mendel's fellow teachers at the *Realschule*, recalled that Mendel suffered a severe blow during his certification examination in Vienna when he engaged in a dispute with the examiner in botany. Although Nowotný did not name the offending examiner, some have since speculated that it must have been Fenzl. This dispute is often recounted, even embellished, in articles about Mendel to claim that he challenged the error of a renowned professor with correct information during his oral examination, creating a predicament so tense that he stormed out of the examination in a huff, collapsing under the weight of the ordeal on his return to the monastery.

But this scenario is unquestionably a myth. If Mendel did not agree with Fenzl's views, which is quite likely, there was no oral confrontation. The dispute was, in fact, between Fenzl and Unger and was well known. As noted earlier, Fenzl argued that plants inherit traits exclusively through the pollen with no contribution from the ovule, so inheritance, in his view, was entirely paternal. Unger, in contrast to Fenzl, claimed that both parents contribute equally to the offspring in plants. Mendel undoubtedly sided with Unger.

The fateful examination, however, was written, not oral. Mendel was isolated in a locked room where he was to compose answers to written questions without access to any source or any person. The only opportunity for Fenzl to challenge Mendel in such a manner would have been in response to the written questions or during the oral *viva voce* phase. According to the May 8, 1856, letter written by Klácel to Bratránek, Mendel abandoned the *written* examination before completing any of it. No oral examination took place. According to the letter,

> P. [Pater] Gregor was called to an examination in Vienna. He left and there was no chance he would be back for the holidays, and since there

> was a shortage for services, I had to stay home. P. Gregor was unlucky. Although he drew easy questions, he fell ill during the first Klausurprüfung [written examination] and as a consequence was unable to write. He seems to have problems with his nerves generally since he endured several such insidious attacks already and they say that in his youth he suffered from epilepsy. The day passed and nothing was achieved. One has to feel sorry for him, since his homework etc. was graded as excellent. But formalities are formalities; in this case it was not possible to continue. Afraid that further attacks might continue, he returned home without accomplishing anything.[8]

On his return to the monastery, Mendel took to his bed with a sickness so overwhelming that his father traveled with haste to attend to him. The monastery ledger lists his father and "brother" staying at the monastery free of charge to care for him. Because Mendel did not have a brother, the "brother" may have been his father's brother, Johann Mendel, who also lived in Heinzendorf, or he may have been Mendel's brother-in-law, Alois Sturm, Veronika's husband, who by then owned the Mendel home. Johann Nowotný, Mendel's fellow teacher at the *Realschule*, recalled that he returned from Vienna sick with "a sore head."[9]

Mendel rescheduled the examination for August 5, 1856, according to university records. However, the monastery kept detailed accounts of Mendel's travel expenses, as did the police in Vienna, and there is no evidence of a trip to Vienna on that date and no university record of Mendel taking the examination. Why he never completed the examination remains a mystery. Although it is often said that Mendel failed his second examination, "abandoned" is a more accurate word. Of the three-part examination, he passed the homework (*eingesandte Arbeiten*), then abandoned and rescheduled the second part (*Klausurprüfung*) but never completed it. The third part (*viva voce*) was not held.

Mendel resumed teaching at the *Realschule* in Brünn, and he continued to do so for twelve more years. His lack of certification was not without consequence; he was forced to continue as a substitute teacher at half salary as a result. Nonetheless, throughout his teaching career, he consistently received accolades from his superiors and high praise from his students.

Mendel's students during this time were young, and several lived into the twentieth century to witness their former teacher achieve worldwide fame. Hugo Iltis interviewed those he could meet in person to record their recollections of the years when Mendel taught them, and others sent him their written recollections. Some skepticism is warranted in that these recollections are seventh-grade memories, recalled decades after their occurrence. As Orel put it, "We should be justified in assuming many of them to have been embellished."[10] Nonetheless, there are common threads, with independent documentation and photographs to support them. Most important, they provide firsthand accounts of how those who personally knew Mendel remembered him.

The single unanimous recollection, borne out not only by verbal recollections but also by official documents and letters, is that Mendel was a superb teacher. According to one student, "He himself delighted so much his work as a teacher, and he was able to present every topic so agreeably and invitingly that we always looked forward to our lessons. . . . Often enough Professor Mendel would deal with some matter jestingly, and thereupon, when his audience laughed heartily, he would blush, join in for a moment in the laughter, and then wave his hand as if to exorcise the spirit he had called up."[11] Another, Joseph Liznar, went on to study mathematics and physics at the University of Vienna and became a professor in geophysics and meteorology at the University of Prague.[12] He recalled, "I had the good fortune of having Gregor Mendel among my teachers. It was he who aroused in me a love for natural science."[13] Mendel did not forget his years of tutoring as a youth. He identified students who required extra help and freely gave them one-on-one supplemental instruction at the monastery. His supervisor reported in 1855 that his teaching was "clear, logical, and well suited to the needs of his hearers."[14] In 1858, Mendel received an official commendation for the quality of his teaching. In it, the *Realschule* director, Joseph Auspitz, noted the collective efforts of teachers at the school to "balance humanistic subjects with the mathematical, scientific and technical subjects" and personally commended Mendel for his "love for science and for the youth that you have exhibited at every opportunity."[15]

Those who knew Mendel also provided accounts of his physical appearance at the time, which are consistent with one another and with photographs: "a man of medium height, broad-shouldered, and already a little corpulent, with a big head and large forehead, his blue eyes twinkling in the friendliest fashion through his gold-rimmed glasses," "brown curls," "curly head, his rather squat figure, his upright gait," "gold-rimmed spectacles," and a "strong Silesian accent."[16] Mendel often shunned clerical clothing. According to one of his students, "Almost always he was dressed, not in a priest's robe, but in the plain clothes proper for a member of the Augustinian order acting as schoolmaster—tall hat; frock-coat, usually rather too big for him; short trousers tucked into top boots."[17] These observations are borne out in two group photographs. The first is of a large tour group traveling to Paris and London in 1862, taken outside a hotel in Paris. In the photo, Mendel is dressed as the other men, in a frock coat, vest, white shirt, and bow tie, with no indication of his clerical status. In another photograph of the faculty of the *Realschule* for the 1864–1865 academic year, he is dressed in a frock coat but wearing a clerical collar, the same as the two other members of his monastery in the photo who were likewise members of the faculty (see figure P2 in this book's prologue).

Several of Mendel's students recalled him tending plants, animals, and bees at the monastery. Outside of class, his students were free to visit him without appointment and often did. According to one, "We all loved Mendel. It was natural, therefore, that from time to time many of us went to visit him at the monastery. We did not wait for an invitation, but arrived like beetles buzzing at an open window, being always received with a friendly smile and taken for a walk in the monastery garden."[18] There his students observed the various plants he cultivated. One recalled him demonstrating the process of artificial hybridization in plants during class, though none recalled the experiments he was conducting at the time. A few accompanied him on botanical excursions to the countryside and to nearby botanical gardens.

The period between Mendel's return from the University of Vienna in late July 1853 and the beginning of his hybridization experiments in the spring of 1856 is important because it marks the time when his thoughts and experiences motivated his classic experiments. The paucity of written records about Mendel during this time has fueled speculation and controversy regarding his motivations. They undoubtedly were powerful; he dedicated an inordinate amount of time and effort to preparing, conducting, presenting, and publishing his epic experiments, consuming more than a decade of his life.

One of Mendel's faculty colleagues and one of his students remembered him keeping animals at the monastery. These recollections have generated a plethora of unfounded speculation, often stated as confirmed fact, that he first observed patterns of inheritance in animals and that this motivated him to study similar patterns of inheritance in plants. Evidence that he raised animals at the monastery, mice in particular, is brief and anecdotal. According to Iltis, "We know from the reports of Hornisch and Nowotný [the former a student, the latter a fellow teacher] that Mendel used to breed mice in his rooms, great mice as well as white mice, crossing these varieties. . . . Mendel himself tells us nothing about this matter, making no reference whatever to his experiments on mice." Iltis then speculated, without evidence, that Mendel shied away from reporting anything about his mouse experiments because "in the eyes of many clericalist zealots it was sufficiently improper for a priest to take any interest in the natural sciences at all, and some persons must have regarded breeding experiments with animals as positively immoral. Mendel had to walk warily for . . . the bishop had a prejudice against him."[19]

Robin Marantz Henig, in her captivating and popular book on Mendel, *The Monk in the Garden*, highlighted his purported mouse-breeding experiments near the beginning of the first chapter, expanding on the rumor Iltis initiated. It is now so widespread that it bears mentioning here, even though it is devoid of evidence. In reference to Bishop Schaffgotsch's 1853–1855 investigation of the monastery, Henig wrote,

> The abbot [Napp] proved too wily an adversary of Schaffgotsch, who was not an especially clever man. But the two clerics did eventually reach a compromise: the monastery could remain open as long as the abbot changed some of the things that Schaffgotsch found most offensive. Among them were the mice that Mendel kept in cages in his two-room flat, where they gave off a distinctive stench of cedar chips, fur, and rodent droppings. He was trying to breed wild-type mice with albinos to see what color coats the hybrids would have. Schaffgotsch seemed to find it inappropriate, and perhaps titillating, for a priest who had taken vows of chastity and celibacy to be encouraging—and watching—rodent sex.
>
> "I turned from animal to plant breeding," Mendel later said with a chuckle. "You see, the bishop did not understand that plants also have sex."[20]

Unfortunately, this account has been cited in respected journal articles and a prize-winning book as evidence that Mendel bred mice and first discovered the principles of inheritance in those experiments but shied away from publishing them because of Bishop Schaffgotsch's scrutiny.[21] University of Kansas zoologist Jack Weir, in addressing Mendel's purported experiments with mice, correctly concluded, "We can have nothing but guess-work about this."[22] What is certain is that Mendel never published anything about inheritance in animals, and his only known experiments with animals were with insects. That Bishop Schaffgotsch would prohibit Mendel from breeding animals is contradicted by the fact that the monastery had large agricultural holdings, which provided substantial income, and breeding agricultural animals was routine. Abbot Napp, in fact, was a member of the sheep breeders association and personally oversaw sheep breeding. Wool production at the monastery's holding was one of its most lucrative enterprises.

Others have suggested that the supposed argument Mendel had with Fenzl when he abandoned his 1856 examination was the principal motivation for his experiments. According to Iltis, "Nowotný believes that this dispute with the examiner led Mendel to begin his experiments, whose origination certainly dates from very soon after the recent defeat."[23] The historical record readily disproves this assertion. As we have seen, the fateful examination was written, not oral, so Fenzl would not have been present. It took place on May 5, 1856, which was somewhat late for pea planting albeit not too late. In all likelihood, however, Mendel's experimental plants were already growing in the garden beds when he traveled to Vienna to take the examination. As an experienced gardener, he would have planted them in late March or early April because peas thrive best in the cool weather of early spring. Moreover, Mendel stated in his classic paper that he had initiated preliminary experiments by cultivating pea varieties as potential parental types over two years to ensure that they bred true-to-type. These preliminary experiments date to the spring–summer seasons of 1854 and 1855, the two years before the purported dispute with Fenzl.

By 1856, Mendel had already been conducting hybridization experiments with a wide range of different plant species, as is evident in the first sentence of his article: "Artificial fertilizations of ornamental plants to produce new color variants led to the experiments to be discussed here."[24] The possibility remains, however, that the dispute between Unger and Fenzl, which Mendel would have known from his time as a university student, was a motivating factor. Mendel's experiments provided definitive evidence supporting Unger's view, a fact that Mendel discussed at length in his article.

A few historians have speculated that the motivation for Mendel's experiments was much more grandiose than clarifying a mere disagreement between his professors. In an effort to cast Mendel as a religious zealot who opposed evolution, they have claimed that he reacted so angrily to reading Darwin's *Origin of Species* that he devised his experiments to disprove Darwin. For example, one author wrote that "Mendel was in favor of the orthodox doctrine of special creation" and that "Mendel's sole objective in writing his *Pisum* paper, published in 1866, was to contribute to the evolution controversy that had been raging since the publication of Darwin's the *Origin of Species* in 1859."[25] Another asserted that "Mendelism came into being historically as a sophisticated form of the doctrine of Special Creation" and that his article "stood in open conflict with the Darwinian conception of evolution as descent with modification by means of Natural Selection."[26]

Serious anachronisms dispel these claims. By the time the first English edition of *Origin of Species* was published in November 1859, Mendel had already completed four spring–summer growing seasons of his hybridization experiments in addition to the two preliminary years of testing the parental varieties in 1854 and 1855. But it is highly unlikely that Mendel read *Origin of Species* before 1863. Mendel did not speak English, and his personal copy, a German translation of the third English edition, has a publication date of 1863 and contains his handwritten annotations.[27] That year, 1863, was the final year of Mendel's experiments, so his reading of *Origin of Species* could not have influenced the planning, design, or conduct of any of his experiments. There is solid evidence, nonetheless, that Mendel relied on *Origin of Species* when writing the manuscript that he would present to the Natural Science Society. He wrote the manuscript for those lectures sometime between the fall of 1863 and the first weeks of 1865.[28]

If Mendel's supposed dispute with Fenzl and his reading of *Origin of Species* were not the inspiration for his experiments, what was? The events in Mendel's life between 1853 and 1856 provide enough pieces of the puzzle to allow construction of a likely scenario. Several topics were hotly disputed during the time Mendel was attending the University of Vienna, and his experiments were designed in such a way that they provided definitive answers to several of these disputes. One was the previously mentioned dispute between Fenzl and Unger on fertilization regarding the relative hereditary contributions of both parents. Another was the role of the environment in shaping inherited characteristics. And among the most important was speculation on the role of hybridization in the evolution of new species, building on more than a century of research.[29] Botanists of the eighteenth and early nineteenth centuries had conducted exhaustive research on hybrids between plant species and how one species might be transformed into another through repeated hybridizations.

Aware of his predecessors' research, Mendel combed through their publications, his aim to build on what they had done. The second paragraph of his article begins with a tribute to them: "Careful observers like *Kölreuter, Gärtner, Herbert, Lecocq, Wichura and others* have tirelessly sacrificed parts of their lives to this objective [plant hybridization]."[30] Mendel's article then singles out Carl Friedrich von Gärtner, whose death in 1850 was relatively recent, his work still fresh in the minds of botanists: "Gärtner especially, in his work 'The Production of Hybrids in the Plant Kingdom,' documented very worthwhile observations."[31] Mendel is referring here to Gärtner's magnum opus, *Experiments and Observations on Hybrid Production in the Plant Kingdom* (*Versuche und Beobachtungen über die Bastarderzeugung im*

Pflanzenreiche), a thick book filled with details of hundreds of experiments on plant hybrids.[32] Mendel cited it eighteen times in his article, more than any other work. Unger was a devotee of Gärtner and probably introduced Mendel to the book. Unfortunately, Gärtner died the year before Mendel began his university studies, so the two never met or corresponded.

On a beautiful summer morning in 1993, in the St. Thomas Monastery, I met with Anna Matalová, a renowned scholar and head of the Mendelianum, a museum devoted to Mendel. With her permission, I browsed through Mendel's personal copy of Gärtner's book, its pages illuminated by the sunlight streaming through a window just a few steps from the place where Mendel conducted his experiments. I quickly noticed how Mendel had annotated page after page with marked passages and handwritten notes. On the inside of the front cover, Mendel had written "Pisum 499," the page where Gärtner recounts his pea experiments—*Pisum* is the Latinized scientific classification of the garden pea. Most significant, however, are Mendel's notes on the page before the back cover, written in dipped ink, referring to characters in pea seeds and plants (figure 4.1).

The notes are technical, referring to characters that vary in the pea plant. Of the several differing pea traits Mendel listed in these notes, he ended up choosing two for his experiments: seed shape and pod shape. According to Robert Olby,

> These notes are important because they show Mendel at work, hunting for clearly marked character differences between the various forms of peas.

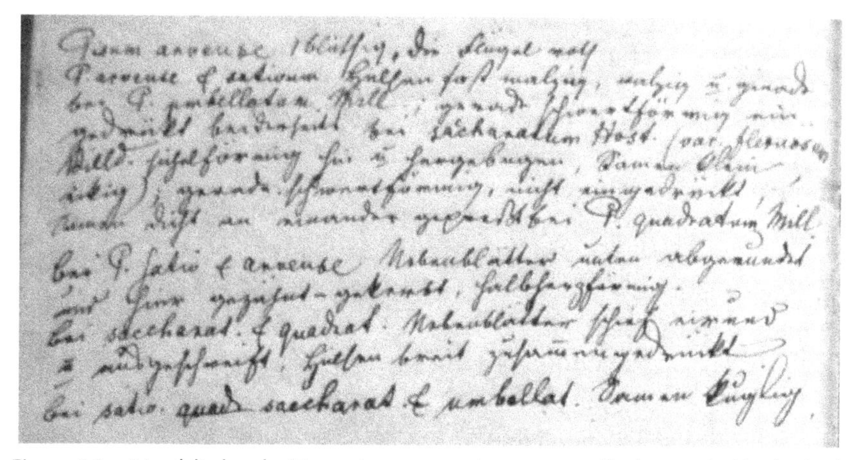

Figure 4.1. Mendel's handwritten notes on pea characters on final page, inside the back cover, of his copy of Gärtner's book *Experiments and Observations on Hybrid Production in the Plant Kingdom. Photograph by Daniel J. Fairbanks.*

> Hence it is reasonable to assume that these notes were written prior to the purchase of the 34 varieties of peas for testing in 1854. If this be so, Mendel must have purchased his copy of Gärtner's book before he had worked out the detailed plan of his experiments.[33]

Not only did Mendel build his research on the experiments of his predecessors, but he was far from alone in studying plant hybrids, even in his own city. The Agricultural Society, of which both Mendel and Napp were members, was focused on applied agricultural science. In 1849, several members organized the Natural Science Section within the society, its purpose to address the pure workings of nature independent of practical applications.[34] Beginning in 1859, the members of this section, Mendel among them, began discussions about expanding the section into an independent society named the Natural Science Society. They formally inaugurated the Natural Science Society in Brünn (*Naturforschender Verein in Brünn*) in December 1861 with Mendel as one of its founding members. Its first meeting was held in January 1862.[35]

At the time of the society's founding, Mendel's pea experiments were in their fifth year. He regularly attended the monthly meetings, where plant hybridization was one of the most frequent topics. Two of Mendel's closest friends, colleagues, and fellow founding members of the society were Gustav Niessl von Mayendorf, an astronomer and botanist, and Alexander Makowsky, a botanist and geologist, both of whom were actively researching plant hybrids. In the first meeting of the society, on January 8, 1862, Niessl lectured on the evolution of new plant species evolving from hybridization. The second meeting, held in February 1862, featured a presentation by a pharmacist and botanist, Carl Theimer, on natural plant hybrids. Later that year, Makowsky published a compilation of plant species in the surrounding region, noting that he had found several natural hybrids, each from two different parental species. The following year, on June 10, 1863, Makowsky reported on observations of plant hybrids in a botanical survey of southern Moravia. Then, on July 8 of that year, Niessl reported his observations of natural plant hybrids in northwest Moravia. At the December 14, 1864, meeting, which was just two months before Mendel would present his pea research, Niessl again reported on plant hybrids observed in nature.[36] Darwin's *Origin of Species* had been discussed in the September 1861 meeting of the Natural Science Section before it became an independent society. Later, in 1862, the society received a copy of the first German edition of *Origin of Species*. Then, as noted in this book's prologue, Makowsky presented a lecture on Darwin's *Origin of*

Species, advocating in favor of Darwin and discussing the role of hybridization in the evolution of new species, the topic of the eighth chapter in Darwin's book. The date of Makowsky's lecture was January 11, 1865. Mendel then presented the next two lectures, in February and March, on his experiments on plant hybrids. Mendel's presentations fit nicely within the context of the society's meetings.

The best source for understanding Mendel's motivation, however, is his own writing. At the beginning of his classic presentation, he stated,

> That a generally standard law for the formation and development of hybrids has not yet been successfully given is no wonder to anyone who knows the extent of the subject and who realises the difficulties with which experiments of this kind must struggle. Some courage is certainly required to undertake such an extensive work; nevertheless, it seems to be the only proper means to finally reach resolution of a question regarding the evolutionary history of organic forms, the importance of which must not be underestimated.[37]

The words "evolutionary history" in this passage are translated from *Entwicklungsgeschichte*, which may alternatively be translated as "evolutionary history" or "developmental history." The context of Mendel's statement makes it clear that the meaning he had in mind is "evolutionary history," which is consistent with Unger's repeated use of the same term in his publications on evolution. Moreover, the only time Darwin wrote of "evolution" in *Origin of Species* was in past-participle form, literally the book's last word: "endless forms most beautiful and most wonderful have been, and are being, evolved."[38] The translator of the German edition of *Origin of Species* that Mendel read and annotated shortly before presenting his experiments employed the word *entwickelt* for Darwin's "evolved," from the same root as *Entwicklung*.

For more than a century, textbook authors have recounted Mendel's objective as determining the fundamental principles of inheritance—particularly the principles of segregation and independent assortment—often taught as Mendel's laws, although the term "principles" is perhaps more accurate. To be sure, he was the first to discover and explain these principles, and he lucidly described them in his article.[39] However, they were not the only question he intended to address. As we will see in the next two chapters, the questions Mendel addressed—and answered—in his classic article fully include but also exceed the so-called Mendelian laws or principles attributed to him in modern textbooks.

Although his discoveries were novel, his research topics were not; they were among the most popular at the time. As Robert Olby succinctly put it, Mendel's work fit "squarely within the context of mid-nineteenth century biology."[40] His experiments, nonetheless, were far more sophisticated than any previous research in plant hybridization. Through his interpretation of the results, he made astounding new discoveries, well beyond what others had accomplished in the past. Most important, he formulated a broad-reaching theory, so well founded that it remains essentially unchanged to this day.

Mendel read his paper to the members of the Natural Science Society in Brünn in two installments, one on the evening of February 8, 1865, and the second a month later, on March 8. It is fortunate for his listeners that he broke the presentation into two lectures. His article (which combines the two lectures into one printed tome) is difficult even for modern scientific readers to digest in a single reading despite its being extraordinarily well organized and lucid. Mendel did not indicate in the article how he divided the text into the two lectures. Fortunately, reporters for local newspapers published brief accounts of each of the meetings shortly after they were held, and the minutes of the March lecture were also published in a local newspaper. From these accounts, it is reasonably apparent where Mendel made the break between the two lectures, and, not surprisingly, the break is logical, dividing the presentations into two major themes.[41]

In the first lecture, he presented his experimental results and the consistent mathematical patterns he observed, including interpretive explanations for those patterns. In the second lecture, he turned his attention to building a broad-reaching theory to explain not only his observations but also those obtained by previous researchers. The next two chapters in this book summarize Mendel's two lectures as he presented them: the first his experiments and the second his theory. For those who wish to read Mendel's original article, the full text is in the appendix at the end of this book. It is an English translation that Scott Abbott, whose expertise is German–English translation, and I published in 2016 to celebrate the sesquicentennial of its original publication.

5 Mendel's First Presentation February 8, 1865
Mathematical Elegance

to finally reach resolution of a question regarding the evolutionary history of organic forms, the importance of which must not be underestimated

—GREGOR MENDEL

MENDEL'S ARTICLE DESCRIBING and interpreting the results of his experiments and his theory of inheritance is a scientific tour de force. To this day, it is one of the most lucid and detailed scientific expositions ever written, a timeless archetype of experimentation. Harvard geneticist Daniel L. Hartl summarized it nicely in an article he and I coauthored:

> Gregor Mendel's celebrated paper is a seemingly inexhaustible source of inspiration and controversy for each succeeding generation of geneticists and historians of genetics. For the aficionado (or the fanatic) it is studied repeatedly, much as an avid sports fan enjoys each rerun of a classic matchup or a movie buff looks forward to yet another screening of *Casablanca*. Mendel's paper is special for a number of reasons. Its historical importance is beyond dispute, but its layout and style are also alluring. Unassuming and unpretentious, Mendel straightforwardly explains his rationale, his experiments, his results, and his interpretation. . . . Above all Mendel's paper appears to reflect the author's simplicity, modesty, and guilelessness.[1]

The article is indeed a timeless classic. As with most classics, readers must carefully study it to grasp the full impact of Mendel's genius. The enormity of information he condensed into his two lectures and his elegant mathematical interpretations were difficult for his hearers and readers to digest in the nineteenth century, when such sophistication was far from the norm. It is no wonder that no one at the time comprehended the significance of his discovery.

The printed article appeared in late 1866 (figure 5.1), more than a year after Mendel gave his presentations.

Mendel wrote in a letter to a colleague that the printed article was "the unchanged reprint of the draft of the lecture mentioned; thus, the brevity of the exposition, as is essential for a public lecture."[2] If we take Mendel at his word, the article is the same as the manuscript he read in the two lectures. This chapter is dedicated to the first lecture, consisting of most of his experiments and his interpretation of them, which he presented on the evening of February 8, 1865. The next chapter centers on the second lecture, wherein he formulated a theory to explain his observations, presented a month later, on March 8, 1865.

According to newspaper accounts, Carl Theimer, vice president of the society, chaired the meeting, opening it with a brief reading of titles for books received for the society's library.[3] He then turned the podium to Mendel, who began by recounting what led him to his experiments and his objective:

> Artificial fertilizations of ornamental plants to produce new color variants led to the experiments discussed here. The striking regularity with which the same hybrid forms reappeared whenever fertilization took place between the same species was the stimulus for further experiments, whose objective was to follow the development of hybrids in their progeny.[4]

Mendel then paid tribute to his forerunners in plant hybridization, especially Gärtner and Kölreuter, followed by a now famous statement, quoted in the previous chapter, that his experiments were "the only proper means to finally reach resolution of a question regarding the evolutionary history of organic forms, the importance of which must not be underestimated."[5] His mention of "the evolutionary history of organic forms" so early in the lecture would not have surprised those in attendance. It was a frequent topic for the society, including the previous month's lecture by Makowsky on Darwin's *Origin of Species*.

Although Mendel viewed his experiments as building on a strong foundation of previous research, he was quick to point out the novelty of his approach in a humble yet confident tone: "Whether the plan by which the individual experiments were arranged and carried out corresponds to the given objective, that may be determined through a benevolent judgment."

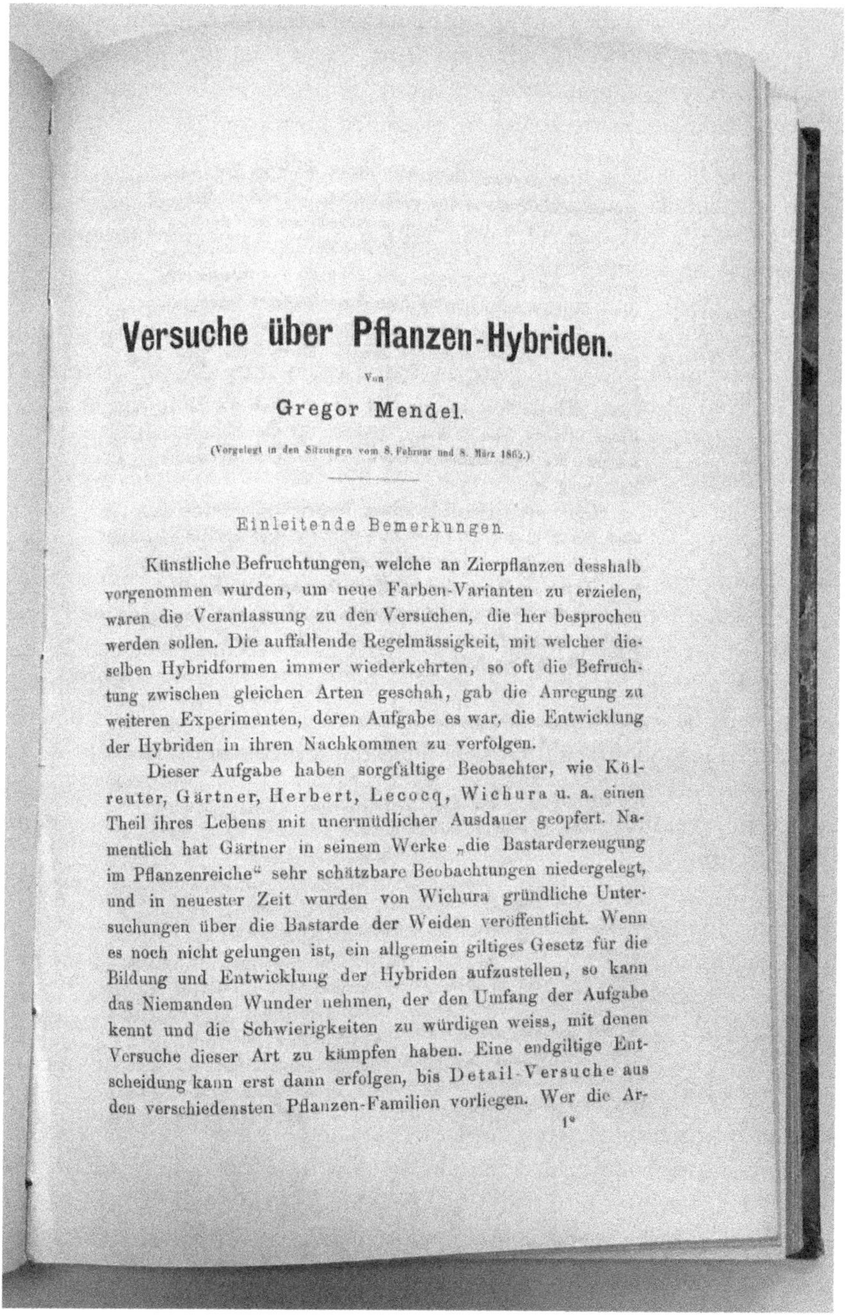

Versuche über Pflanzen-Hybriden.

Von

Gregor Mendel.

(Vorgelegt in den Sitzungen vom 8. Februar und 8. März 1865.)

Einleitende Bemerkungen.

Künstliche Befruchtungen, welche an Zierpflanzen desshalb vorgenommen wurden, um neue Farben-Varianten zu erzielen, waren die Veranlassung zu den Versuchen, die hier besprochen werden sollen. Die auffallende Regelmässigkeit, mit welcher dieselben Hybridformen immer wiederkehrten, so oft die Befruchtung zwischen gleichen Arten geschah, gab die Anregung zu weiteren Experimenten, deren Aufgabe es war, die Entwicklung der Hybriden in ihren Nachkommen zu verfolgen.

Dieser Aufgabe haben sorgfältige Beobachter, wie Kölreuter, Gärtner, Herbert, Lecocq, Wichura u. a. einen Theil ihres Lebens mit unermüdlicher Ausdauer geopfert. Namentlich hat Gärtner in seinem Werke „die Bastarderzeugung im Pflanzenreiche" sehr schätzbare Beobachtungen niedergelegt, und in neuester Zeit wurden von Wichura gründliche Untersuchungen über die Bastarde der Weiden veröffentlicht. Wenn es noch nicht gelungen ist, ein allgemein giltiges Gesetz für die Bildung und Entwicklung der Hybriden aufzustellen, so kann das Niemanden Wunder nehmen, der den Umfang der Aufgabe kennt und die Schwierigkeiten zu würdigen weiss, mit denen Versuche dieser Art zu kämpfen haben. Eine endgiltige Entscheidung kann erst dann erfolgen, bis Detail-Versuche aus den verschiedensten Pflanzen-Familien vorliegen. Wer die Ar-

1*

Figure 5.1. The first page of Mendel's printed article, published in 1866. *Photograph by Daniel J. Fairbanks.*

Mendel selected the common garden pea as his experimental organism. Referring to it by its Latin scientific name, *Pisum*, and as a member of the legume family, Leguminosae (the family of pea, bean, soybean, peanut, clover, alfalfa, and locust trees), he explained his reasoning:

> From the beginning, special attention was given to the *Leguminosae* because of their curious floral structure. Experiments made with several members of this family led to the conclusion that the genus *Pisum* sufficiently meets the necessary requirements.[6]

These "necessary requirements" include the fact that the pea plant is naturally true-breeding, meaning that all the offspring of a naturally self-fertilized plant are identical to one another and to their parent plant. Mendel, in fact, showed experimentally and mathematically *why* this is the case, verifying the scientific accuracy of the common saying "as alike as two peas in a pod." All plants grown from seeds harvested from the same pod, as well as those from the same plant, are genetically identical to their parent plant and to one another, like an innumerable proliferation of multiple identical twins, generation after generation. The word Mendel used to refer to the consistent identical nature of such offspring is *constant.*

The exception, so important to Mendel's work, is the recent offspring of experimentally hybridized plants of the type Mendel generated with his hands through artificial cross-pollination. To describe the offspring of these artificially produced hybrids, he used the term *variable.* Absent such human intervention, the pea plant naturally self-fertilizes to produce constant offspring.

Mendel noted that pea plants "possess uniform characters that are easily and certainly distinguishable, and they give rise to perfectly fertile progeny when reciprocally crossed." After considering several of these characters that are "certainly distinguishable," he narrowed his experiments to seven pairs of them.

In 2016, at the sesquicentennial of Mendel's article, I had the opportunity to unveil my oil paintings of these seven pairs of characters in the Moravian Museum in Brno, not far from where he grew his plants. These paintings now reside in the museum's collection, where they are on exhibit. The descriptions in figure 5.2 are Mendel's own words, which succinctly define these character-pairs, with images of my paintings illustrating his descriptions.

The seven character-pairs Mendel included in his pea experiments. Mendel's descriptions accompany each image.

1. The difference in the form of the ripe seeds. These are either spherical or somewhat rounded . . .; or they are irregularly angular and deeply wrinkled.

2. The difference in the color of the seed albumen. . . . The albumen [interior] of the ripe seeds is pale yellow, bright yellow, or orange colored; or it possesses a more or less intensive green color.

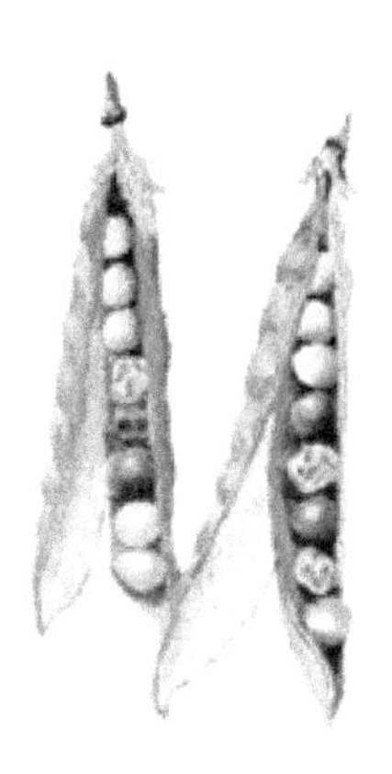

3. The difference in the color of the seed coat. This is either colored white, a character consistently associated with white flower color, or it is grey, grey-brown, or leather brown with or without violet spots, in which case the color of the standard petal appears violet, that of the wings purple, and the stem at the base of the leaf axils is tinged reddish. The grey seed coats turn blackish brown in boiling water.

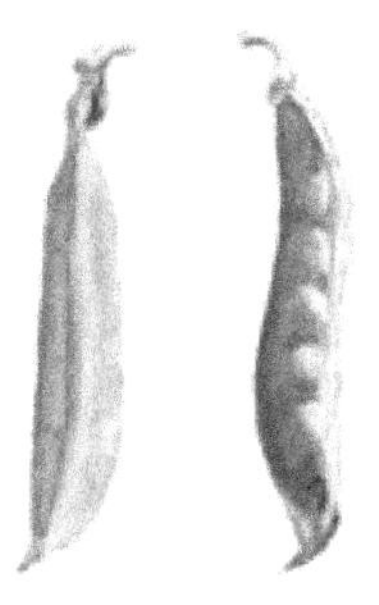

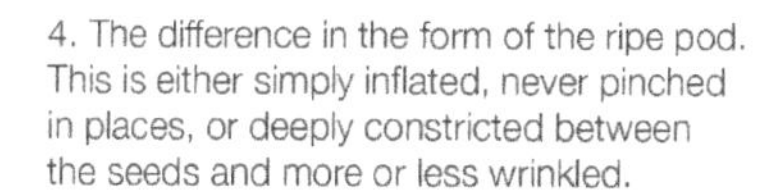

4. The difference in the form of the ripe pod. This is either simply inflated, never pinched in places, or deeply constricted between the seeds and more or less wrinkled.

5. The difference in the color of the unripe pod. It is either light to dark green or colored a bright yellow.

Figure 5.2. The seven character-pairs Mendel included in his pea experiments. Mendel's descriptions accompany each image. *Paintings by Daniel J. Fairbanks, oil on panel. Collection of the Mendelianum, Moravian Museum, Brno.*

6. The difference in the placement of the flowers. They are either axial, i.e., distributed along the stem, or terminal, accumulated at the end of the stem.

7. The difference in the length of the stem. The length of the stem is very different in individual forms; however, for each one it is a constant character undergoing insignificant changes insofar as the plants are healthy and are raised in the same soil. In the experiments with this character, to obtain a confident difference, the long stem of 6–7 feet was united with the short one of 0.75–1.5 feet.

Several of these contrasting character-pairs were readily available to Mendel in common varieties of peas, and the same is true today. For example, round seeds are starchy, used for soups and stews, such as split-pea soup. Wrinkled seeds are sugary and are picked early during their growth while the pods are still succulent for fresh and frozen peas. The sugar within the tender, immature peas conveys a sweet taste. Yellow and green seeds are obvious in packages of dried peas for yellow or green split-pea soup. Shelling varieties for frozen packaged peas have inflated pods that are tough and stringy but make it easier for people (or today machines) to split open the still-green pods to remove the succulent peas without damaging them. Edible-pod peas, by contrast, like those for stir-fry recipes, have constricted pods that adhere to the seeds, and they are tender with no stringiness, allowing them to easily be eaten whole. They are more difficult to open than inflated pods. Tall pea plants climb on trellises, whereas dwarf peas can be grown without support. Most commercial pea varieties have white flowers because the colored seed coats of purple-flowered types tend to darken during cooking, as mentioned by Mendel. Some heirloom varieties with colored seed coats have lovely colored flowers with various shades ranging from dark purple to pale pink.

Mendel had considered other characters, mentioned in his presentation, such as pod size, length of the peduncle (the "stem" that connects the flower and pod to the plant), and time of flowering. However, he excluded them because

> some of these characters, however, do not permit certain and sharp separation because the difference rests on a "more or less" that is difficult to determine. Such characters could not be used for the individual experiments, which had to be limited to characters that appear clearly and decidedly in the plants.[7]

By limiting his experiments to these seven character-pairs, he could unambiguously distinguish the two contrasting types from each other, allowing him to protect against misclassification. Seed shape and seed color at times can be difficult to distinguish with absolute certainty, as Mendel noted, writing that the mature and dry seeds "often vary in color and shape, but with some practice in sorting, errors are easily prevented."[8] He could absolutely distinguish the five contrasting characters in the plants (as opposed to those in the seeds) with no chance of error. For example, the tall plants were considerably taller (over six feet as Mendel described them) than the dwarf plants (at most about a foot and a half), the difference clear and unmistakable.

Mendel designed his first experiments to ensure that the two parents in each cross differed by just one character-pair and were the same for the other six; in his words, "plants were used that differed in only one essential character."[9] This approach makes good scientific sense. By focusing on one pair of characters in each experiment, Mendel excluded any confounding effects that might arise if the inheritance of one character-pair influenced the inheritance of another.

Mendel further stated that for each experiment, he hybridized pairs of plants in both directions, with one type as the female parent and the other as the male parent, and vice versa. He called these "reciprocal crosses" (*wechselseitige Kreuzung*). By doing so, he experimentally tested the dispute between his professors, Unger and Fenzl. If Unger was correct and both parents contribute equally to the offspring, the hybrids should appear the same regardless of whether the female or the male carried a particular character. If Fenzl was correct and the pollen parent contributes all the inherited material to the offspring, the offspring should always bear the character of the male parent, and the hybrids derived from reciprocal crosses should differ. Mendel's observations unambiguously confirmed Unger's proposition. In Mendel's words, "It has been shown through all the experiments that it is completely unimportant whether the dominant character belongs to the seed plant or to the pollen plant; the hybrid form remains exactly the same in both cases."[10]

For each experiment, he allowed the flowers he had fertilized to fully mature into pods that eventually dried and contained fully developed dry seeds. He called these seeds and the plants grown from them the *hybrid generation*, which modern biologists and textbook authors designate as the *F_1 generation*. Because this difference between Mendel's designations and modern designations can be confusing, I'll use Mendel's designation first, then the modern designation in parentheses.

After the plants had matured and dried, Mendel collected the hybrid seeds from the pods that arose from the flowers he had artificially fertilized. For two of the character-pairs—seed shape and seed color—he could observe the results by midsummer, directly in the hybrid (F_1) seeds that arose from the flowers he cross-pollinated, due to the nature of how pea seeds develop. He had to postpone characterizing the remaining five character-pairs until he planted the seeds and observed the plants that grew from them in the next spring season. He chose to conduct the largest experiments with the seed-shape and seed-color character-pairs because these experiments required considerably less garden space than the other experiments, and he could tabulate the results at the end of each growing season when he harvested the seeds.

In the seed-shape experiment, Mendel noted that all the hybrid (F_1) seeds were round, like their round-seeded parent and unlike their wrinkled-seeded parent. He observed a similar pattern in the seed color experiment: all hybrid (F_1) seeds had yellow interiors like their yellow-seeded parent and unlike their green-seeded parent.

To observe the remaining five character-pairs, Mendel planted the hybrid (F_1) seeds the following spring to observe the hybrid (F_1) plants. The pattern for them remained the same as for the seed character-pairs: all displayed one of the two parental characters and not the other. Mendel co-opted two common terms to describe this phenomenon as it applies to inheritance. He referred to the parental character that appeared in the hybrid (F_1) seeds and plants as "dominant" (*dominirende*) and the one that did not as "recessive" (*recessive*). In his words,

> Each of the seven hybrid characters either resembles one of the two original parental characters so perfectly that the other one escapes observation or is so like it that a confident distinction cannot be made. This circumstance is of great importance for the determination and classification of the forms appearing among the progeny of the hybrids. In the following discussion those characters that are transmitted wholly or nearly unchanged in the hybrid association, that themselves represent the hybrid characters, are defined as *dominant*, and those that become latent in the association are defined as *recessive*. The term "recessive" was chosen because the so-named characters recede or completely disappear in the hybrid, but among the progeny thereof, as is shown later, reappear unchanged.[11]

He then allowed the hybrid (F_1) plants to naturally self-fertilize to produce what he called the "first generation of the hybrids," which is what biologists now call the F_2 generation. Some authors who are not familiar with the pea plant have mistakenly assumed that Mendel intercrossed the hybrid plants to obtain this next generation. Doing so would have been extraordinarily laborious and time consuming—and completely unnecessary. All he had to do at this stage was plant the seeds and allow the plants to naturally self-fertilize.

At this point, Mendel's presentation becomes increasingly mathematical, building on his simpler experiments to explain those that are more complex. Despite his clarity and logic, readers often become lost in the numbers, a curse that most certainly afflicted those attending his lecture. Likewise, as you read the rest of this chapter, I urge you to avoid the temptation to stop as the numerical complexity increases because, in the end, the mathematical elegance of Mendel's discovery will become apparent in his explanation of consistent and beautiful patterns.

Also obvious in Mendel's presentation is his skill as a teacher. As excellent teachers often do, he tells the listeners what to look for before he presents the numbers:

> In this generation [F_2], *along with the dominant* characters, the *recessive* characters reappear in their full individuality and do so in the determinate and pronounced average ratio of 3:1, so that of every four plants from this generation, three produce the dominant and one the recessive character. This applies without exception for all characters included in the experiment.[12]

Having stated the pattern to watch for—three dominant individuals for every recessive one in every experiment—he then presents the detailed results of the first two experiments on seed shape and seed color, which were by far the two largest experiments. In his words,

> The ratios acquired for each pair of two differing characters are as follows:
>
> - *First experiment*: Shape of the seeds. From 253 hybrids, 7324 seeds were obtained in the second experimental year. Of these seeds 5474 were round or somewhat rounded, and 1850 were angular wrinkled. The resulting ratio is 2.96:1.
> - *Second experiment*: Color of the albumen. A total of 258 plants produced 8023 seeds, 6022 yellow and 2001 green; the former relate to the latter in the ratio 3.01:1.[13]

After explaining some details about these first two experiments, he turns his attention to the remaining five experiments on character-pairs observable in plants:

> - *Third experiment*: Color of the seed coat. Of 929 plants, 705 produced violet-red flowers and grey-brown seed coats; 224 had white flowers and white seed coats. This results in a ratio of 3.16:1.
> - *Fourth experiment*: Shape of the pods. Of 1181 plants, 882 had simply inflated and 299 had constricted pods. Hence the ratio is 2.95:1.
> - *Fifth experiment*: Color of the unripe pod. The number of experimental plants was 580, of which 428 had green and 152 had yellow pods. Thus the ratio of the former to the latter is 2.82:1.

- *Sixth experiment*: Position of the flowers. Of 868 cases, the flowers were located along the stem 651 times and were terminal 207 times. This ratio is 3.14:1.
- *Seventh experiment*: Length of the stem. Of 1064 plants, 787 had long stems, and 277 had short ones. Hence this relative ratio is 2.84:1.[14]

He now repeats the straightforward conclusion he had previously asked the reader to notice: "If the results of all experiments are summarized, there is an average ratio between the number of forms with dominant and recessive characters of 2.98:1 or 3:1."[15]

Here is where Mendel surpassed his predecessors in his choice to examine large numbers of offspring and carefully tabulate those numbers in search of a consistent pattern. The experiments in every case revealed the same mathematical ratio that no one before him had recognized: *the dominant character appeared in three-fourths of the offspring of the hybrids and the recessive character in one-fourth.*

He then carried all seven experiments into the next generation, allowing the plants to naturally self-fertilize. He called this the "second generation of the hybrids," now known as the F_3 generation. He observed that "those forms that preserve the recessive character in the first generation [F_2] do not vary in the second generation [F_3] in relation to that character; they remain *constant* in their progeny."[16] In other words, all the offspring of plants with a recessive character also display that recessive character; every individual is identical to its parent. This is, however, not true for the dominant character. Mendel again tells the reader what to expect before presenting his results:

> This is not the case for those that possess the dominant character in the first generation [F_2]. Of these *two-thirds* yield progeny that carry the dominant and recessive character in the ratio 3:1 and thus show the same behavior as the hybrid forms; only *one-third* remains constant with the dominant character.[17]

He then presents the numerical data he collected to document this conclusion. He begins, as before, with the seed-shape and seed-color experiments, which again were the largest:

> The individual experiments produced the following results:

- *First experiment*: Of 565 plants raised from round seeds of the first generation, 193 produced only round seeds and thus remained constant in this character; 372, however, simultaneously produced round and angular seeds in the ratio 3:1. Thus the number of hybrid types relative to the number of constant types is 1.93:1.

- *Second experiment*: Of 519 plants raised from seeds whose albumen in the first generation had the yellow color, 166 produced exclusively yellow; 353, however, produced yellow and green seeds in the ratio 3:1. This resulted in division of hybrid and constant forms in the ratio 2.13:1.[18]

Although the five experiments with plant characters were smaller because of limitations for garden space, the mathematical patterns he observed were the same. In Mendel's words,

> For each of the following experiments, 100 plants were selected that retained the dominant character in the first generation, and to test its signification, 10 seeds from each were cultivated.

- *Third experiment*: The progeny of 36 plants produced exclusively grey-brown seed coats; from 64 plants some with grey-brown and some with white seed coats were produced.
- *Fourth experiment*: The progeny of 29 plants had only simply inflated pods; of 71, however, some had inflated and some had constricted pods.
- *Fifth experiment*: The progeny of 40 plants had only green pods; from those of 60 plants some had green and some had yellow pods.
- *Sixth experiment*: The progeny of 33 plants had flowers located only along the stem; of 67, however, some had flowers located along the stem, and some had terminal flowers.
- *Seventh experiment*: The progeny of 28 plants produced long stems; from 72 plants some had long stems and some had short stems.[19]

As a careful scientist, Mendel noted that the results of the fifth experiment (pod color) were not as close to the 2:1 ratio he expected. To be certain that this observation was due to a random-chance deviation and not something unusual about this character-pair, he repeated this experiment. The second time, the results were closer to what he expected, which allowed him to draw a consistent conclusion from the pattern evident in all the experiments:

> The fifth experiment, which showed the largest deviation, was repeated and then, instead of the ratio 60:40, produced the ratio 65:35. *The average ratio 2:1 consequently appears certain*. Thus it is proved that of each form possessing the dominant character in the first generation, two-thirds carry the hybrid character; one-third, however, remains constant with the dominant character.[20]

At this point, Mendel is ready to venture into an abstract mathematical representation to explain the consistent pattern he observed. In his words,

> If A represents one of the two constant characters, for example the dominant, a the recessive, and Aa the hybrid form in which the two are united, then the expression
>
> $$A + 2Aa + a$$
>
> shows the developmental series for the progeny of the hybrids of each pair of divergent characters.[21]

In other words, if the underlying characters are defined as A and a, then the 3:1 ratio breaks down into a symmetrical 1:2:1 ratio distributed as $A + 2Aa + a$. This expression is straightforward combinatorial mathematics of the type he had studied in detail under his physics and mathematics professor, Andreas von Ettingshausen, while at the University of Vienna.

Had he ended his experiments and his interpretations here, his discovery would have been momentous. In fact, some authors end their discussion of Mendel's experiments at this point (or before it) for the sake of simplicity. Mendel, however, was far from finished, and his application of combinatorial mathematics was about to literally increase exponentially.

His next step was to experiment with the very situation he had excluded in the first set of experiments: determining whether the inheritance of one character-pair has any influence on the inheritance of another. As he put it, "The next objective consisted of researching whether the developmental law found for each pair of differing characters was valid when several different characters are united in the hybrid through fertilization."[22]

He reported detailed results for one experiment combining two character-pairs (seed shape and seed color) and one experiment combining three character-pairs (seed shape, seed color, and seed coat color). After reporting the results of these two experiments, he simply stated that "various other experiments were undertaken with a smaller number of experimental plants in which the rest of the characters were associated in twos and threes in the hybrids; all produced approximately the same results."[23]

The experiment with two character-pairs was from a cross between a variety with yellow-round seeds and one with green-wrinkled seeds. As expected, all the hybrid seeds were yellow and round, the two dominant characters he had observed in his previous experiments. The results in the first generation of the hybrids (F_2) were the following:

Yellow round	315
Yellow wrinkled	105
Green round	108
Green wrinkled	32

If the inheritance of one character-pair has no influence on the inheritance of another, Mendel predicted that he should observe two 3:1 ratios independently superimposed on each other, one for seed shape and one for seed color, which mathematically is the ratio 9:3:3:1. This breaks down into a distribution of 9 yellow round seeds, 3 yellow wrinkled seeds, 3 green round seeds, and 1 green wrinkled seed. His results were very close to this 9:3:3:1 ratio, deviating from a perfect result by just a few individuals in each category.

Mendel then mathematically expands this ratio to impose two 1:2:1 ratios independently on each other to predict which proportions of each type of seed should breed true and which proportions should bear different types of offspring. Letting *A* and *a* represent the underlying inherited elements that confer round and wrinkled seeds and *B* and *b* the elements that confer yellow and green seeds, he tabulated the following results, using the progeny from the second generation of the hybrids (F_3) to determine the underlying constitution of their parents:[24]

38	*AB*
35	*Ab*
28	*aB*
30	*ab*
65	*ABb*
68	*aBb*
60	*AaB*
67	*Aab*
138	*AaBb*

In this case, he compared the observed results to his prediction of two 1:2:1 ratios independently superimposed on each other, which is the ascending ratio of 1:1:1:1:2:2:2:2:4, which is how Mendel depicted it. However, like the 1:2:1 ratio, it can also be expressed symmetrically as 1:1:2:2:4:2:2:1:1. The observed results were again exceptionally close to this overall ratio he predicted.

The largest experiment Mendel conducted was with three character-pairs—seed shape, seed color, and seed coat color—which he referred to as the experiment that "required the most time and effort."[25] He carried this experiment through the second generation from the hybrids (F_3) and reported the following results:

8 *ABC*	22 *ABCc*	45 *ABbCc*
14 *ABc*	17 *AbCc*	36 *aBbCc*
9 *AbC*	25 *aBCc*	38 *AaBCc*
11 *Abc*	20 *abCc*	40 *AabCc*
8 *aBC*	15 *ABbC*	49 *AaBbC*
10 *aBc*	18 *ABbc*	48 *AaBbc*
10 *abC*	19 *aBbC*	
7 *abc*	24 *aBbc*	
	14 *AaBC*	78 *AaBbCc*
	18 *AaBc*	
	20 *AabC*	
	16 *Aabc*	

In this case, he predicted an ascending ratio of 1:1:1:1:1:1:1:1:2:2:2:2:2:2:2:2:2:2:2:2:2:4:4:4:4:4:4:8, which, like the earlier ratios, can be rearranged symmetrically as 1:1:1:1:2:2:2:2:2:2:4:4:4:8:4:4:4:2:2:2:2:2:2:1:1:1:1. As before, the observed results were very close to those he predicted.[26]

At this point, Mendel draws one of the most important conclusions of his experiments, one that he emphasizes through profuse use of italics:

> There is, then, no doubt that for all of the characters admitted into the experiments the following sentence is valid: *The progeny of hybrids in which several essentially differing characters are united represent the terms of a combination series in which the developmental series for each pair of differing characters are combined.* Simultaneously it thus is shown that *the behavior of each pair of differing characters in hybrid association is independent of other differences between the two original parental plants.*[27]

As he consistently observed in each these experiments, the inheritance of each character-pair is entirely independent of the inheritance of the others.

After some clarification of this emphatic point, Mendel ended his first lecture. It undoubtedly left at least some the audience members lost with mathematical tabulations and complex ratios swimming in their brains. Nonetheless, it was a brilliant statistical analysis of the type previously unseen in biological studies, revealing a consistent set of elegantly symmetrical patterns that had been evident in plain sight throughout human history but never before deciphered.

In his next lecture, a month later, Mendel interpreted his observations to develop an expansive theory. He was unaware that his theory would explain the mysteries of reproduction, inheritance, and the evolution of new species—not just in pea plants but in most species on earth, including humans.

Mendel's Second Presentation, March 8, 1865 — 6
A Momentous Theory

a constant law that is founded in the material nature and arrangement of the elements

—GREGOR MENDEL

THE MARCH 8 MONTHLY MEETING of the Natural Science Society in Brünn was devoted to the second part of Mendel's two-part presentation. As in the previous month, Vice President Carl Theimer chaired the meeting. After reading the list of books and specimens in the mineral collection that had been donated to the society during the preceding month, he turned the podium over to Mendel.

In this lecture, Mendel derived a momentous theory to explain his results. I use the word "theory" here in its true scientific context, which is much stronger than the usual understanding of the word. According to the National Academy of Sciences (USA), the scientific definition of "theory" is "a well-substantiated explanation of some aspect of the natural world that can incorporate facts, laws, inferences, and tested hypotheses."[1] In every respect, Mendel's theory fully accords with this definition. Scientific theories typically require modification, refinement, and expansion over time as new evidence accumulates, and Mendel's theory is no exception. Nonetheless, it has remained as one of the most durable theories in the history of science, a tribute to his ability to design experiments and accurately interpret the results.

At the beginning of this second lecture, before explaining his theory, Mendel still had some experimental results to present. Reserving them for this lecture made sense because they directly apply to the derivation of his theory. In his printed article, this initial part of the lecture is in a section titled "The Fertilizing Cells of the Hybrids," where he introduced the first part of his theory: the union of two cells, one from each parent, to form one foundational cell at fertilization.

These experiments were a spin-off of an experiment he had presented in the previous lecture. In that earlier experiment, his objective was to determine whether the inheritance of two character-pairs is independent. As a reminder, he predicted and observed a 1:1:2:4:2:1:1 ratio, confirming complete independence for inheritance of two character-pairs.

Now, in this second lecture, Mendel explains how he conducted the same hybridization as before, between a round-yellow seeded variety and a wrinkled-green seeded one, to obtain round-yellow hybrid seeds. These he planted. But instead of relying on natural self-fertilization for the next generation, he crossed these hybrid plants back to both parents, a process known as backcrossing. He did so reciprocally so that the hybrid served as the female parent with each of the two parental types as the male parent and also the hybrid as the male parent with each of the two parental varieties as the female parent. This resulted in all four possible combinations of backcrosses. He then collected the seeds derived from these backcrosses, planted them, and allowed them to naturally self-fertilize. If his previous deductions were correct, there should be four possible outcomes in each of the experiments, and each should be equally likely, resulting in a 1:1:1:1 ratio in each of the four experiments. As predicted, the results in each case were close to a 1:1:1:1 ratio.

Now, as a final confirmation, he presented a slightly different experiment, this time with flower color and stem length as the two character-pairs. He hybridized a short-stemmed variety that had colored flowers (recessive and dominant) with a long-stemmed one that had white flowers (dominant and recessive). As expected, the hybrid offspring consisted entirely of plants with long stems and colored flowers. These he crossed not to their parents but instead to a short-stemmed variety with white flowers (recessive and recessive). Again, according to his earlier interpretation of independent inheritance, he predicted a 1:1:1:1 ratio, which, in fact, was the case.

With the results of these experiments and those presented in the first lecture, he now had ample evidence to devise a theory to explain how underlying hereditary units and the cells harboring them were partitioned into reproductive cells and how they behaved during fertilization. He named these reproductive cells "germ cells" and "pollen cells," which correspond, respectively, to "egg cells" and "sperm cells" in animals and humans.

He explained how these reproductive cells form in mathematical terms for any one character-pair in hybrids:

> In their formation, pollen and germ cells of the forms A and a occur in equal proportions on average in fertilization, and thus each form appears twice, since four individuals are formed. Therefore, participating in fertilization are the pollen cells, $A + A + a + a$; and the germ cells, $A + A + a + a$.[2]

Then, if fertilization is a truly random event, there are four possible ways for pollen and germ cells to unite, each equally likely, which Mendel presented with the following diagram:[3]

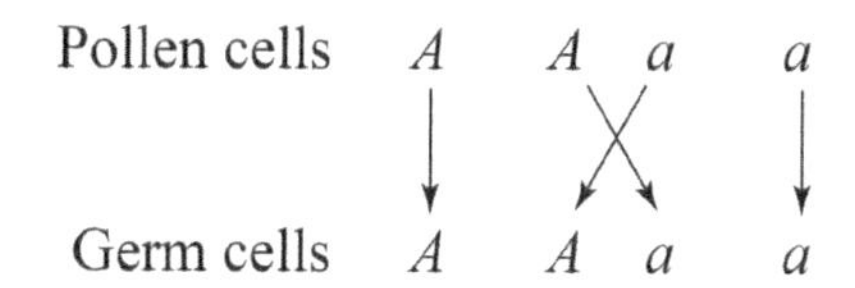

From this explanation, he then deduced one of the key points of this theory. In his words,

> The result of fertilizations can be clearly illustrated if the designations for united germ and pollen cells are shown as fractions, with the pollen cells above the line, the germ cells below. Thus, in this case,[4]

$$\frac{A}{A}+\frac{A}{a}+\frac{a}{A}+\frac{a}{a}$$

This straightforward expression is the foundation of Mendel's theory, and it is one of the most important leaps in scientific history. First, Mendel identifies what we now call a *gene* and its *variants*, representing the gene as a letter and its variants as upper- and lowercase versions of that letter, such as A and a. In this case, A and a are partitioned into the germ and pollen cells so that each cell carries either A or a but not both. Second, he portrays the random union of these cells during fertilization, which results in a distinct mathematical pattern of inheritance consisting of all four possible combinations in equal proportions. This expression fully explains his observations from the first lecture for each of the seven character-pairs. The first three combinations, A/A, A/a, and a/A display the dominant character, and only a/a displays the recessive character, consistent with Mendel's repeated observations of a 3:1 ratio in his experiments. As Mendel pointed out, this expression also breaks down into the 1:2:1 ratio he deduced and confirmed experimentally:

This accounts for the striking phenomenon that the hybrids are able, like the two original parental forms, to produce progeny that are identical to themselves; A/a and a/A both produce the same combination Aa, because, as alluded to earlier, it makes no difference for the result of fertilization which of the two characters belongs to the pollen or germ cells. Thus,[5]

$$\frac{A}{A}+\frac{A}{a}+\frac{a}{A}+\frac{a}{a}=A+2Aa+a$$

At this point in his presentation, Mendel clarifies why the results he observed *approximated* but did not exactly match the ratios he predicted and why it was essential for him to design experiments with large numbers of individuals:

> This is the *average* course for the self-fertilization of hybrids when two differing characters are united in them. In individual flowers and in individual plants, the condition through which members of the series are formed, however, can undergo alterations that are not insignificant. . . . It remains wholly left to chance which of the two kinds of pollen fertilizes each individual germ cell. Thus the individual values necessarily undergo fluctuations and even extreme cases are possible. . . . The true numerical ratios can be derived only as the mean from the sum of the largest possible number of individual values; the larger their number is, the more mere chance effects are eliminated.[6]

Mendel then turned to the case of two character-pairs combined (A and a plus B and b), which results in the expansion of his previous equation into its sixteen possible combinations, likewise in equal proportions, as he depicted it:[7]

$$\frac{AB}{AB}+\frac{AB}{Ab}+\frac{AB}{aB}+\frac{AB}{ab}+\frac{Ab}{AB}+\frac{Ab}{Ab}+\frac{Ab}{aB}+\frac{Ab}{ab}+\frac{aB}{AB}+$$
$$\frac{aB}{Ab}+\frac{aB}{aB}+\frac{aB}{ab}+\frac{ab}{AB}+\frac{ab}{Ab}+\frac{ab}{aB}+\frac{ab}{ab}$$

His experimental observations for two character-pairs fully conformed to these mathematical expressions, the observed values deviating only slightly from them.

Turning to three character-pairs, he refrained from providing the exponentially expanded mathematical expression required for this situation (sixty-four possible combinations) and simply stated,

> The developmental series for hybrids can be accounted for in a quite similar manner when *three kinds of differing characters* are combined. The hybrid forms eight different types of germ and pollen cells *ABC*, *ABc*, *AbC*, *Abc*, *aBC*, *aBc*, *abC*, *abc* and once again each pollen type unites on average once with each germ cell type.[8]

An obvious question for Mendel was whether the mathematical patterns he had discovered in the garden pea are unique to this species or whether they apply broadly to other plant species (he made no mention of animals or humans). He told his listeners that he had conducted preliminary experiments with the common bean on a small scale and that the patterns conformed with those in his pea experiments. He also addressed in this section an important topic that Darwin had raised in *Origin of Species*, although he did not mention Darwin by name.

After his presentations, Mendel continued to experiment with other plant species for another eight years. During this time, he acquired and read other books by Darwin, experimentally testing Darwin's suppositions. I will defer this section of his lecture to the next chapter because the topics he briefly introduced here are those he continued to investigate in the years to come.

Having shown in preliminary experiments with the common bean that the patterns of inheritance he observed in the garden pea were also evident in another species, he now was ready to draw some sweeping conclusions in a section he titled "Concluding Remarks," which historian and current head of the Mendelianum Jiří Sekerák characterized as "the most valuable part of Mendel's achievement."[9] Here Mendel ventures into one of the most groundbreaking and innovative aspects of his theory—what scientists now call "the theory of the gene."

The word "gene" did not exist at the time; a Swedish scientist, Wilhelm Johannsen, coined it in 1909. Instead, Mendel adopted a word he had not yet mentioned until this point, apparently saving this word for his conclusion. He had throughout his lectures employed the words "character" (*Merkmal* in German) and "factor" (*Factor* in German) to denote inherited character-pairs. Now he introduced the word "elements" (always in the plural, *Elemente* or *Elementen* in German), and his use of this word differed entirely from its use to designate chemical elements. According to his definition in the context of inheritance, "elements" signified not the outward visible traits, like seed shape or seed color, but rather *the inherited units within the cells* that confer those outward traits. His use of this term in this manner was not novel. Mendel's botany professor, Franz Unger, had used the same word in the same context in his book *Botanical Letters*, as had Gärtner in his book on plant hybrids and Darwin in *Origin of Species*, books Mendel had studied.

There was much speculation at the time about the substance of inheritance. Some imagined a microscopic individual encapsulated within each pollen grain (or sperm cell in animals and humans), a notion called "preformation." Were this correct, inheritance would be purely paternal, as Mendel's professor Eduard Fenzl stubbornly asserted until his death. Notably, Makoswky discussed this notion of pure paternal inheritance in his January 1865 presentation on Darwin and dismissed it as incorrect. Mendel, in his lectures, made a giant leap, presenting abundant experimental evidence to definitively show *why* it was incorrect.

In Mendel's day and later, some claimed that inheritance was beyond the material world, a function of some immaterial essence that could never be identified, isolated, or purified. Others held the view that the inherited substance was material in nature, present within both female and male reproductive cells, and inherited from both parents when those cells united at fertilization. As Mendel was about to show, this was the only explanation fully consistent with his observations.

Mendel's first use of the term "elements" is in one of the most revealing passages of his lectures, providing the foundation of his theory:

> According to the view of famous physiologists, in phanerogams [seed-bearing plants], for the purpose of reproduction, one germ cell and one pollen cell unite into a single cell that is able to develop into an independent organism through the uptake of matter and the formation of new cells. This development takes place according to a constant law that is founded in the material nature and arrangement of the elements, which succeeds in a viable union in the cell.[10]

His phrase "material nature and arrangement of the elements" (in his original German *der materiellen Beschaffenheit und Anordnung der Elemente*) is uniquely powerful. Here, Mendel clarifies that the elements of heredity are material units, which arrange themselves "according to a constant law." Sekerák has focused on this phrase as one of the most important of Mendel's inferences, suggesting the translations "material composition" or "material structure" as alternatives to "material nature."[11] Mendel refers to "famous physiologists," a statement that undoubtedly is a reference to Unger, in making the argument that "one germ cell and one pollen cell unite into a single cell that is able to develop into an independent organism." This process of fertilization—the union of germ and pollen cells in plants and the union of egg and sperm cells in animals and humans—was by no means universally accepted at the time. Mendel's results provided indisputable evidence of it. To reaffirm the magnitude of this line of reasoning, Mendel added a long footnote to this passage, again using the word "elements," wherein he utterly dismantles Fenzl's view and fully supports Unger's (without naming either):

> With *Pisum* [pea] it is shown without doubt that there must be a complete union of the elements of both fertilizing cells for the formation of the new embryo. How could one otherwise explain that among the progeny of hybrids both original forms reappear in equal number and with all their peculiarities? If the influence of the germ cell on the pollen cell were only external, if it were given only the role of a nurse, then the result of every artificial fertilization could be only that the developed hybrid was exclusively like the pollen plant or was very similar to it. In no manner have experiments until now confirmed that. Fundamental evidence for the complete union of the contents of both cells lies in the universally confirmed experience that it is unimportant for the form of the hybrid which of the original forms was the seed or the pollen plant.[12]

Rarely is Mendel acknowledged as the one who discovered the first solid evidence that two reproductive cells, one from the female parent and one from the male, unite to form a single cell that gives rise to each individual. As we will see in the next chapter, Darwin continued to argue that parental equality in fertilization was *not* the case, and he did so *after* Mendel's article had been published, although he was unaware of it.[13] Only as abundant evidence accumulated by the end of the nineteenth century did scientists come to fully accept this view.

After powerfully showing how fertilization must ensue, Mendel then presents what is now widely known as his *law of segregation* (sometimes named as "Mendel's first law"), taught in essentially every biology class. His argument is logical and straightforward, again employing the term "elements" to denote what we now call "genes":

> In relation to those hybrids whose progeny are *variable*, one might perhaps assume that there is an intervention between the differing elements of the germ and pollen cells so that the formation of a cell as the foundation of the hybrid becomes possible; however, the counterbalance of opposing elements is only temporary and does not extend beyond the life of the hybrid plant. Because no changes are perceptible in the general appearance of the plant throughout the vegetative period, we must further infer that the differing elements succeed in emerging from their compulsory association

> only during development of the reproductive cells. In the formation of these cells, all existing elements act in a completely free and uniform arrangement in which only the differing ones reciprocally segregate themselves. In this manner the production of as many germ and pollen cells would be allowed as there are combinations of formative elements.[14]

This is one of the most extraordinary passages of Mendel's entire article, bursting with information compressed into a single paragraph. To explain its meaning, allow me to depict it pictorially. First, Mendel refers to the "differing elements of the germ and pollen cells" that form "a cell as the foundation of the hybrid." If we use Mendel's symbols, *A* and *a*, to denote these "differing elements," then the two unite to form "the foundational cell of the hybrid" (what modern biologists call a zygote):

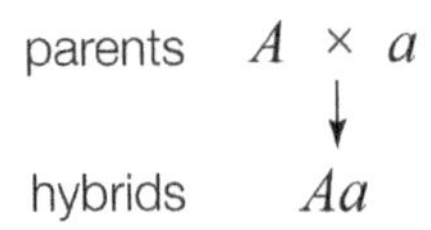

Mendel perceived this union of "opposing elements" as a "counterbalance," perceptively noting that *A* and *a* do not blend into each other but remain distinct. This counterbalance of *Aa* persists "throughout the vegetative period." Then "the differing elements succeed in emerging from their compulsory association only during development of the reproductive cells," a process now known as meiosis. In other words, the differing elements remain associated as *Aa* until those differing elements segregate from each other in the reproductive organs (which, in the case of the pea plant are within the flowers) into individual germ cells and pollen cells:

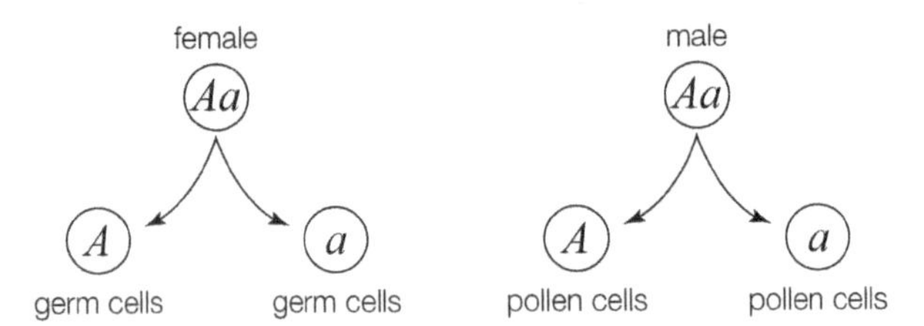

With self-fertilization, there are four possible ways for germ and pollen cells to unite, which Mendel had previously depicted as a series of fractions where the numerators represent the elements from the pollen cells and the denominators the elements from the germ cells:

$$\frac{A}{A}+\frac{A}{a}+\frac{a}{A}+\frac{a}{a}$$

In the final sentence of this paragraph, Mendel succinctly states what is now known as his *law of independent assortment*: "In this manner the production of as many germ and pollen cells would be allowed as there are combinations of formative elements." Shortly thereafter, he reiterates this law of independent assortment: "the behavior of each pair of differing characters in hybrid union is independent of the other differences between the two original plants and, further, . . . the hybrid produces as many types of germ and pollen cells as there are possible constant combination forms."[15] In other words, all possible combinations of elements appear in the fertilizing cells in equal proportions, explaining the exponentially expanded results of his experiments in which two or more character-pairs are combined.

Mendel was not content to allow his work to stand on its own but felt compelled, perhaps even honored, to extend his theory to explain the work of his predecessors. In his words, "It may not be without interest to compare the observations made herein on *Pisum* with the results of successful research by the two authorities in this area, Kölreuter and Gärtner."[16]

Here, he notes that his predecessors had observed two types of hybrids. In some plant species, hybrids may appear as intermediates between their parents, then produce offspring that are uniformly like themselves when self-fertilized, their offspring retaining the intermediate hybrid appearance generation after generation. Mendel called this type a "constant hybrid" because, when a hybrid naturally self-fertilizes, its offspring do not vary from it or from one another throughout subsequent generations. Hybrid plants in other groups of species, however, when self-fertilized produce offspring that *vary* in their characteristics when compared to their hybrid parent. Mendel called this type a "variable hybrid" because individual members of the offspring vary in their characteristics in subsequent generations. The pea plant clearly falls into the variable-hybrid category, as Mendel understood well before he began his experiments. As he presented his lectures, he was already planning experiments on several plant species that he suspected would produce constant hybrids. Four years later, in 1869, he would present a lecture on constant hybrids to the Natural Science Society.

Neither Mendel nor anyone else at the time knew how broad reaching his theory was. The principles of inheritance he discovered in the pea plant were the same as those in most species. Beyond the mathematical patterns he documented, he also clearly articulated his deduction that underlying material elements conferred outwardly visible inherited traits, referring to the material nature of those elements.[17] Four years later, in 1869, a young Swiss scientist named Johann Friedrich Miescher was the first to isolate DNA from cells.[18] Twentieth-century scientists would later show that the material hereditary elements Mendel described were genes in DNA, confirming the ingenuity of Mendel's deductions.

Mendel would eventually gain fame as one of the history's most extraordinary scientists—the founder of genetics. Arguably, in hindsight, Darwin and Mendel were the two most influential biologists of the nineteenth century. Well into the twentieth century, their theories would be fully synthesized into a single theory, sometimes called neo-Darwinism, as the foundation of modern biology. Although Darwin experienced worldwide fame and notoriety in his day, the same cannot be said of Mendel. Despite the exhaustive nature of his research, his insightful interpretation of it, the clarity of his exposition, and his derivation of a theory that has endured for more than a century and a half, no one (including Mendel himself) recognized its importance during his lifetime.

Mendel and Darwin 7

The laws governing inheritance are quite unknown.

—CHARLES DARWIN

MENDEL'S THEORY OF INHERITANCE was the most crucial missing puzzle piece for Darwin's theory of evolution by natural selection—a piece that, though readily available, Darwin never found. Just a few pages into the first chapter of *Origin of Species*, Darwin begins a paragraph with a very telling sentence: "The laws governing inheritance are quite unknown."[1] Implied in this sentence and repeatedly expounded throughout the book is Darwin's recognition that there are, in fact, laws governing inheritance, that they are important to his theory, and that they were "quite unknown." Darwin was unaware that, in 1859 when *Origin of Species* was published, Mendel was in the fourth year of hybridization experiments, and his theory would reveal those laws.

When Darwin published the fifth edition of *Origin of Species* in 1869, he changed this sentence to reflect his understanding of how little things had improved: "The laws governing inheritance are for the most part unknown."[2] Such a statement exemplifies the obscurity of Mendel's theory at the time, for he had correctly formulated the laws of inheritance in a well-written and lucid article that was available to Darwin in libraries in England. However, even for those who heard Mendel's lectures or read his article, the connection with Darwin's theory remained elusive.

In the years that elapsed between the publication of Mendel's article until his theory was rediscovered in 1900, scientists would debate various hypotheses of heredity as ways of explaining Darwinian evolution by descent through modification and natural selection. Foremost among them was Darwin himself, who proposed a hypothesis that failed when tested experimentally. Why Mendel's discovery was absent from these debates when it was accessible remains a perplexing question—one with several possible answers.

A likely reason for the neglect is that no one, including Mendel, had yet recognized the importance of his discovery or its broad applicability to plants, animals, and humans. In the demeanor of a cautious scientist reluctant to prematurely draw sweeping conclusions, Mendel warned that the interpretation of his experiments could not yet be considered as generally applicable, cautioning in his classic article, with his emphasis in italics, that "a final determination will result only when *detailed experiments* on the most diverse plant families are available."[3] Mendel's focus remained only on plants, not animals or humans. Although he hoped others would conduct these "detailed experiments," he was not content to leave that task to them. Before completing his research with peas, he had already started similar hybridization experiments with other plant species belonging to diverse families to determine whether the patterns he had discovered were generalizable.

Mendel and Darwin's paths drew physically close in 1862, while Mendel's experiments were in their penultimate year. Mendel signed up for a grand railroad and steamship tour beginning in Vienna and proceeding through Paris to London. He is in the center of a photograph with more than 150 members of the tour, taken in Paris (figure 7.1). By then, *Origin of Species* had been in print in English for three years. However, Mendel could not speak or read English, and the German translation he owned, read, and annotated was published in 1863, the year after his tour. For these reasons, he probably knew little if anything about Darwin at the time. At one point on the trip, he passed within just a few miles of Darwin's home in Downe, probably unaware of it, and records of Darwin's correspondence show that he was at home while Mendel was nearby.[4]

In an ill-timed irony, the excitement and controversy surrounding *Origin of Species* may have eclipsed Mendel's discovery. By the time Mendel presented his experiments and theory in 1865, Darwinism had spread well beyond England. Mendel and most of his academic colleagues in both Vienna and Brünn, including the other friars in his monastery, were well acquainted with Darwin's book by the time Mendel presented his lectures.

As I mentioned in this book's prologue and in chapter 4, Alexander Makowsky presented a lecture on Darwin's *Origin of Species* in the January 1865 session of the Natural Science Society in Brünn. The published text

Figure 7.1. Close-up of Gregor Mendel in a photograph of members of a large tour group traveling to Paris and London in 1862. The entire photograph (not shown) is of a group of more than 150 people standing in front of a hotel in Paris. Note that Mendel's clothing is similar to that of other men in the photo, with no indication of his status as a priest. *Period photograph, Mendelianum, Moravian Museum, Brno, public domain.*

of Makowsky's lecture, which appeared in the same printed volume as Mendel's article, shows remarkable concordance with Mendel's lectures.[5]

For example, both Makowsky and Mendel mention how difficult it is to distinguish species from varieties, a topic Darwin repeatedly confronted in *Origin of Species*. Makowsky wrote,

> A definite line has certainly not been drawn, neither between species and subspecies, nor between subspecies and varieties, nor finally between slight varieties and individual differences. Arranged in a row, they indistinguishably overlap and establish the perception of a true transition. On the basis of these and other examples, Darwin concludes that the terms: species and variety are arbitrarily chosen, and do not permit a precise distinction.[6]

Mendel stated in his February lecture, "It is as impossible to draw a sharp line of distinction between species and varieties as it is to establish a fundamental distinction between the hybrids of species and varieties."[7] He annotated several passages in his personal copy of *Origin of Species* on this subject, one of which reads (in Darwin's original English),

> From these remarks it will be seen that I look at the term species, as one arbitrarily given for the sake of convenience to a set of individuals closely resembling each other, and that it does not essentially differ from the term variety, which is given to less distinct and more fluctuating forms. The

> term variety, again, in comparison with mere individual differences, is also applied arbitrarily, and for mere convenience sake.[8]

Both Makowsky and Mendel addressed the dichotomy that Darwin raised throughout his book, whether species had been specially created or had arisen through descent with modification. According to Makowsky, "As much as this [mutability of species] may stand in contradiction to the past, it has at least the same legitimacy as the contrary view that species are immutable products of nature."[9] And from Mendel,

> Through the success of transformation experiments, Gärtner was persuaded to oppose the opinion of those naturalists who dispute the stability of plant species and assume continuous evolution of plant species. He sees in the completed transformation of one species into the other the unambiguous evidence that a species has fixed limits beyond which it cannot change. Although this view cannot be afforded unconditional validity, nonetheless a confirmation deserving notice regarding the supposition made earlier about the variability of cultivated plants is found in the experiments performed by Gärtner.[10]

Although there is no direct evidence that Mendel conferred with Makowsky regarding their sequential presentations in 1865, these and other similarities in their sequential presentations suggest that they may have done so or that Mendel may have adapted his presentation to follow up on some of the topics addressed by Makowsky. Mendel took an interest in Makowsky's lecture as evidenced by his annotations in his copy of it.[11] In any case, Mendel's presentations logically followed Makowsky's, fully compatible with what was then widespread interest in Darwinism.

Given the 1863 publication date for Mendel's personal copy of *Origin of Species*, it obviously did not influence his design or execution of his experiments. But there is strong evidence that it impacted his thinking as he prepared his lectures and wrote his article. The passages he annotated in his copy of Darwin's book provide ample evidence of his interest in Darwin's writings.[12] Not surprisingly, the chapter with the most marked passages is chapter 8, titled "Hybridism," which dealt with the role of hybridization in the evolution of new varieties and ultimately new species, a subject Mendel addressed in his own research.

One passage on page 302 in Mendel's copy of *Origin of Species* apparently intrigued him more than any other, as indicated by his handwritten note "pag. 302" on the book's first page and the double-lined mark on page 302 (figure 7.2). In Darwin's original English, this passage reads,

"The slight degree of variability in hybrids from the first cross or in the first generation, in contrast with their extreme variability in the succeeding generations, is a curious fact and deserves attention."[13] Mendel undoubtedly recognized this phenomenon as one he repeatedly observed in his experiments: the uniformity of hybrids and the variability of their offspring in successive generations. And he had discovered the hereditary mechanism that fully explains this "curious fact."

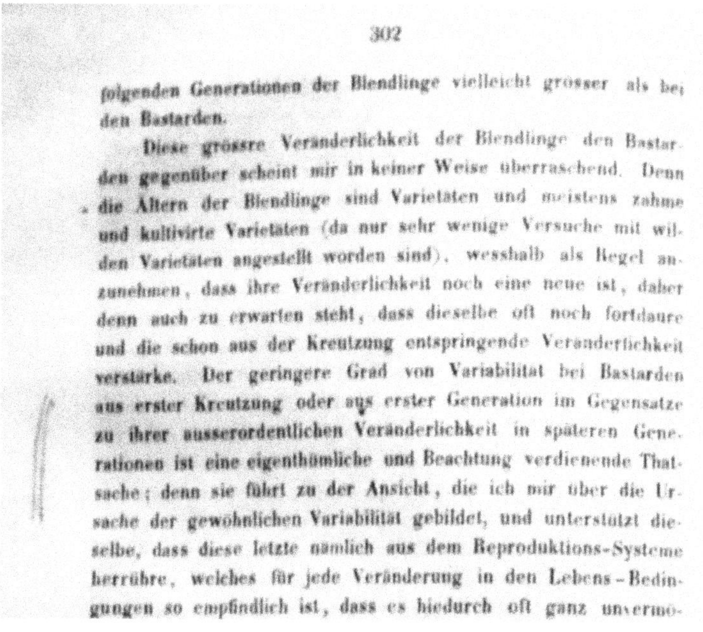

302

folgenden Generationen der Blendlinge vielleicht grösser als bei den Bastarden.

Diese grössere Veränderlichkeit der Blendlinge den Bastarden gegenüber scheint mir in keiner Weise überraschend. Denn die Ältern der Blendlinge sind Varietäten und meistens zahme und kultivirte Varietäten (da nur sehr wenige Versuche mit wilden Varietäten angestellt worden sind), wesshalb als Regel anzunehmen, dass ihre Veränderlichkeit noch eine neue ist, daher denn auch zu erwarten steht, dass dieselbe oft noch fortdaure und die schon aus der Kreutzung entspringende Veränderlichkeit verstärke. Der geringere Grad von Variabilität bei Bastarden aus erster Kreutzung oder aus erster Generation im Gegensatze zu ihrer ausserordentlichen Veränderlichkeit in späteren Generationen ist eine eigenthümliche und Beachtung verdienende Thatsache; denn sie führt zu der Ansicht, die ich mir über die Ursache der gewöhnlichen Variabilität gebildet, und unterstützt dieselbe, dass diese letzte nämlich aus dem Reproduktions-Systeme herrühre, welches für jede Veränderung in den Lebens-Bedingungen so empfindlich ist, dass es hiedurch oft ganz unvermö-

Figure 7.2. Page 302 in Mendel's personal copy of *Origin of Species* with his double-line annotation. The marked passage reads, in Darwin's original English, "The slight degree of variability in hybrids from the first cross or in the first generation, in contrast with their extreme variability in the succeeding generations, is a curious fact and deserves attention." *Photograph by Daniel J. Fairbanks.*

Had Mendel presented his discovery as a mechanism that explains Darwin's theory, it might have been more widely recognized, but Mendel never mentioned Darwin in his article. Even so, there are enough passages in it with Darwinian wording to show that his study of *Origin of Species* had a strong influence on his writing.[14] One of the most Darwinian of the terms Mendel used is *Lebensbedingungen*, which, in Darwin's original English, is "conditions of life." Darwin repeatedly employed this term more than 100 times throughout *Origin of Species*. In his article, Mendel twice used the German term *Lebensbedingungen*, which he encountered often in his German translation of *Origin of Species*, for Darwin's "conditions of life" (notice *Lebens-Bedingungen* in Darwin's book in the final two lines in figure 7.2). The following passage shows how Mendel employed this term in his classic article:

> No one will seriously assert that the development of plants in a natural landscape is governed by different laws than in a garden bed. Here, just as there, typical variations must appear if the conditions of life [*Lebens-bedingungen*] are changed for a species, and it has the ability to adapt to the new conditions. It is freely admitted, through cultivation the production of new varieties is favored, and by the hand of man many a variation is preserved that would have failed in the wild state, but nothing gives us the right to assume that the tendency for new varieties to form is so extremely augmented that species soon lose all stability and that their offspring break up into an infinite array of highly variable forms. If the change in the conditions of vegetation were the sole cause of variability, then one would be justified in expecting that those domesticated plants cultivated under almost the same conditions for centuries would have acquired stability. As is well known, this is not the case, for especially among them not only the most different but also the most variable forms are found.[15]

The famed British statistician, mathematician, and geneticist Sir Ronald Fisher wrote in 1936 about this passage:

> The reflection of Darwin's thought is unmistakable, and Mendel's comment is extremely pertinent, though it seems to have been overlooked. He may at this time have read the *Origin*, but the point under discussion may equally have reached his notice at second hand.[16]

Fisher recognized the obvious Darwinian nature of this passage, though his comment reveals that he was unaware that Mendel owned a copy of the *Origin of Species* and had read it. Mendel marked three passages with the term *Lebensbedingungen* in his copy of *Origin of Species*. One is in the first passage he marked in the book, in chapter 1, and it is directly related to his assertion. From Darwin's original English,

> It seems pretty clear that organic beings must be exposed during several generations to the new conditions of life to cause any appreciable amount of variation; and that when the organization has once begun to vary, it generally continues to vary for many generations.[17]

This also was a topic Makowsky pointed out in his presentation: "If we look at our oldest cultivated plants and domestic animals, we notice above all that the individuals of a species differ considerably more from one another than the individuals of a species in nature, which variations we might regard as the result of widely varying conditions of life."[18]

German scholar Scott Abbott and I, in 2016 at the sesquicentennial of Mendel's classic paper, published a new English translation of it, taking a unique approach to this translation. Based on the premise that Mendel was reading *Origin of Species* at the time he was writing his manuscript and that this was probably the only work he was reading at the time that was originally written in English, we searched for German words in common in Mendel's paper and the German translation of *Origin of Species* that he was reading. We then sought out the English words used by Darwin for these German word matches and employed them as much as possible, making it a "Darwinized translation" with a distinct nineteenth-century flavor.

We also color coded the words to distinguish those in passages Mendel annotated in his copy of *Origin of Species* from those that are not in annotated passages but are found elsewhere in the book, suspecting that Mendel may have referred to the passages he marked when writing his manuscript. What we discovered was astounding. Words in the passages Mendel marked were noticeably clustered toward the end of his article in the final two sections (from his second lecture in March) and were especially prevalent in one paragraph, the same one to which Fisher alluded when he wrote "the reflection of Darwin's thought is unmistakable." We published an article summarizing the abundant evidence we found of Darwin's influence on Mendel, which is freely available online.[19]

As discussed in the previous chapter, another Darwinian term in Mendel's paper that stood out because of how and where Mendel employed it is the German word *Elemente*, which is directly translatable as the English word "elements." Notably, Darwin employs the word "elements" in *Origin of Species* in the same manner as Mendel to refer to the inward hereditary material. For example, in reference to fertilization, Darwin wrote in *Origin of Species*, "the two sexual elements which go to form the embryo."[20] Unger also used the German word *Elemente* in the same manner and in an evolutionary context in his book *Botanical Letters*, which was published while Mendel was enrolled in his classes at the University of Vienna, here in English translation: "who can deny that new combinations of the elements arise out of this permutation of vegetation, ever reducible to a certain law—combinations which emancipate themselves from the preceding characteristic of the species and appear as new species?"[21]

There is no question that Mendel was fully aware of Darwin's work when he presented and published his paper. Even so, he did not mention

Darwin by name at this point in his writings. He would mention Darwin in a subsequent publication four years later and several times in letters after he had read additional books by Darwin. Yet there is no evidence that Mendel made any attempt to contact Darwin. And, despite many efforts, scholars have found no evidence that Darwin was aware of Mendel's work.

How close did Darwin come to learning of Mendel? When Mendel became famous in the early twentieth century, years after both he and Darwin had died, Darwin's son Francis searched his father's collection (by then removed from Down House and in Francis's possession) for any evidence that Darwin had Mendel's writings. He found none. There was no reprint and no copy of the journal containing Mendel's article.

Mendel had forty reprints of his article available to send to other scientists, and the fate of only a few is known.[22] A common rumor in the mythology of Mendel is the unfounded claim that Darwin owned a reprint of Mendel's paper but that it was uncut and therefore unread. In fact, the rumor is even more expansive, a presumption that Mendel sent uncut reprints to others and that most of the recipients didn't take the time to cut the reprints and read them.[23] One author went so far as to assert that the "more common story" of uncut reprints "is a beautiful metaphor for the one thing we know for sure—that most of the scientists who received Mendel's reprint never bothered to read it."[24]

There is, however, strong contrary evidence that the reprints were, in fact, cut before Mendel sent them and that at least two recipients read them. Definitive evidence that the reprints were cut comes from several typesetter errors that Mendel corrected by hand before sending the reprints, which he could do only after the reprints had been cut. For example, he made a unique correction nine pages into each one. On this page, a sentence, as Mendel intended it, should begin with the words "*Je zwei*," meaning "Each two." However, the reprints contain a typesetter's error, reading "*In zwei*," affecting two adjacent letters. Mendel took a pen and changed the uppercase "I" to a "J" by adding a curve to the bottom of the "I" and even drawing a small spherical dot at the end of the curve to resemble a "J" in printed font. He then crossed out the "n" and wrote an "e" above it (figure 7.3). This unique correction is present in reprints sent to different recipients, indicating that Mendel must have made the corrections before sending them rather than the recipients making them. He could make corrections such as this in each reprint only if it had been cut.

Nonetheless, there is no evidence that Mendel sent Darwin a reprint, cut or uncut. Mendel's name, nonetheless, was in Darwin's possession; two books in his personal library referenced Mendel, both in German.

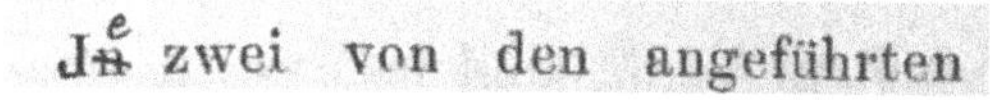

J~~n~~e zwei von den angeführten

Figure 7.3. Mendel's hand correction of the typesetter error "In zwei" to "Je zwei" on page 9 of the reprints of his article. *Photograph by Daniel J. Fairbanks.*

The earlier of these two books is one by Hermann Hoffmann, published in 1869, titled *Researches on Determining the Value of Species and Varieties: A Contribution to the Critique of the Darwinian Hypothesis* (*Untersuchungen zur Bestimmung des Werthes von Species und Varietät: Ein Beitrag zur Kritik der Darwin'schen Hypothese*).[25] Hoffmann made several brief references to Mendel's classic article, and there is evidence that Darwin viewed a page with Mendel's name on it but probably paid it no attention.

The page with Mendel's name that Darwin most certainly saw is page 52. A long footnote occupies most of this page, and Hoffmann mentions Mendel three sentences into the footnote, referring to Mendel's experiments with beans (not peas), stating (in English translation),

> According to Mendel, hybrids can be obtained from *P. vulgaris* + *nanus*, from *Multiflorus* (mas) + *nanus* (fem.). The hybrids of *mult.* + *nanus* were not very fertile or even completely infertile. The color of the flowers and seeds was mostly closer to *multifl.* In the second generation, several plants arose which stood closer to *nanus*.[26]

Darwin, because of his interest in bean hybrids, made handwritten notes on pages 50, 51, 53, and 54 of Hoffmann's book, so he undoubtedly saw page 52, though he might have paid no attention to Mendel's name because it was a footnote in small-print Gothic German, embedded in a set of references to several authors. Mendel is the fifth of eight authors referenced in this footnote. Even if Darwin had read it, the brief information Hoffmann chose to include on that page would not likely have led Darwin to Mendel's article.

Hoffmann also made brief references to Mendel's paper on pages 86, 112, 119, and 136. The reference on page 136 is the only one that mentions Mendel's pea experiments. It too is very brief and only discusses the fact that pea flowers naturally self-fertilize, that they can be easily cross-fertilized by hand, and that the pea weevil might on occasion bring in foreign pollen when it enters the flower. Had Darwin read this passage, he would have learned nothing about Mendel's experiments or his theory, nor would he have learned anything he did not already know about pea plants. Darwin made no notes on or near this page.

Perhaps more important was the book *The Plant Hybrids* (*Die Pflanzen-Mischlinge*) by Wilhelm Olbers Focke,[27] which contains short summaries of Mendel's research in peas and beans. Although brief, these summaries contain enough details that they might have led Darwin to Mendel were it not for the timing. Focke's book was published in 1881, the year before Darwin died. Darwin received an advance copy in November 1880 while he was immersed in preparing his final book, *The Formation of Vegetable Mould through the Action of Worms, with Observations on Their Habits*.[28] Shortly after receiving the book by Focke, Darwin loaned it to a colleague, writing, "I will send by today's post a large book by Focke, received a week or two ago, on Hybrids, and which I have not had the time to look at."[29]

Not long thereafter, on April 19, 1882, Darwin passed away. It is no surprise, therefore, that the pages in Darwin's copy of Focke's book containing summaries of Mendel's research remain uncut to this day and therefore unread.[30] The discovery of these uncut pages in a book owned by Darwin probably morphed into the rumor about Darwin's supposed uncut reprint of Mendel's paper.

How might Mendel's research have influenced Darwin had he known of it? There is no shortage of speculation on this question, and historians have arrived at conclusions ranging from one extreme to the other. Some surmised that Darwin would have embraced Mendel's work, recognizing how well it explained the elusive "laws governing inheritance," which were so crucial to his theory of natural selection. Others claim that the two espoused such disparate views that Darwin would have paid little attention to Mendel's research. Meager circumstantial evidence can be found to support both of these opposing views as well as a range of views lying between them.

There is no dispute that Darwin was intensely interested in the mechanism of heredity and its relationship to hybridization. Also clear in Darwin's writings is his frustration over the lack of a workable theory of inheritance. He published an ill-fated hypothesis on the mechanism of heredity that he named "The Provisional Hypothesis of Pangenesis," the title of a chapter in his two-volume work *The Variation of Animals and Plants under Domestication*.[31] According to Darwin's hypothesis, various anatomical parts of organisms release hereditary particles, which he called "gemmules," and they purportedly circulate in the blood of animals and in the sap of plants and then coalesce in the reproductive organs to be transmitted to the next generation. The different types of gemmules, as Darwin hypothesized, mixed with one another during cross-fertilization to produce hybrids that were, for the most part, intermediates between their parents.

The prediction that this sort of inheritance should produce intermediate hybrids was a point of contention in Darwin's and Mendel's day. That contention would escalate in disputes regarding inheritance throughout their lifetimes and into the early years of the twentieth century, well after both had died. Mendel had observed absolute and complete dominance for all seven character-pairs he researched. He made it clear, however, that dominance was not universal and that he had personally observed intermediate hybrids for certain characters. But he viewed precise intermediacy as the exception rather than the rule, as is evident in the following passage from his classic article:

> The experiments conducted with ornamental plants in past years already produced evidence that hybrids, as a rule, do not represent the precise intermediate form between the original parents. With individual characters that are particularly noticeable, like those related to the form and size of the leaves, to the pubescence of the individual parts, the intermediate form is in fact almost always apparent; in other cases, however, one of the two original parental characters possesses such an overwhelming dominance that it is difficult or quite impossible to find the other in the hybrid.
>
> Such is exactly the behaviour of the Pisum [pea] hybrids. Each of the seven hybrid characters resembles one of the two original parental characters either so perfectly that the other one escapes observation or is so like it that a confident distinction cannot be made. This circumstance is of great importance for the determination and classification of the forms appearing among the progeny of the hybrids.[32]

Darwin, by contrast, viewed intermediacy as the rule and supposed that dominance is an anomaly that tends to produce sterility in hybrids, as the following passage from *Origin of Species* illustrates:

> So again amongst hybrids which are usually intermediate in structure between their parents, exceptional and abnormal individuals sometimes are born, which closely resemble one of their pure parents; and these hybrids are almost always utterly sterile, even when the other hybrids raised from seed from the same capsule have a considerable degree of fertility.[33]

Darwin's contrary view did not escape Mendel's notice. He marked this passage in his German copy of *Origin of Species*.

Recognizing that pangenesis was merely a hypothesis with little experimental support, Darwin wrote,

> I give my well-abused hypothesis of Pangenesis. An unverified hypothesis is of little or no value; but if anyone should hereafter be led to make

observations by which some such hypothesis could be established, I shall have done good service, as an astonishing number of isolated facts can be thus connected together and rendered intelligible.[34]

Darwin's "good service" exceeded his expectations. Pangenesis became a widespread working hypothesis of inheritance throughout the latter part of the nineteenth century, probably because of the high regard scientists throughout the world held for Darwin. It also has been a favorite target of iconoclasts, giddy to wave the banner that "Darwin was wrong!"

Rarely, however, is Darwin recognized, as he should be, for the eventual demise of pangenesis. According to Darwin's hypothesis, gemmules circulated in the blood of animals. Therefore, it should be possible to transfer gemmules from one animal to another through blood transfusions, and, if pangenesis is correct, they should be inherited in the offspring of transfusion recipients. Darwin's half cousin, Francis Galton, conducted a series of blood-transfusion experiments in rabbits to test the hypothesis, hoping to find evidence of transferred gemmules. However, no such evidence emerged despite large-scale transfusions of blood from one rabbit to another. While Galton was willing to accept the results, Darwin was stubbornly reluctant. But after a time, he acquiesced. In a letter, Emma Darwin (his spouse) wrote that the "experiments about rabbits are failing, which is a dreadful disappointment to them both [Galton and Darwin]."[35]

In the meantime, Mendel had a German translation of both volumes of Darwin's *The Variation of Animals and Plants under Domestication*, and he read the second volume carefully, as evidenced by his fifty-seven annotations.[36] As mentioned in this book's prologue, on one of the pages, Mendel underlined passages in which Darwin explained pangenesis, emphatically made a large exclamation point in the margin, and then wrote on the bottom of the page, "*sich einem Eindrucke ohne Reflexion hingeben* [to indulge in an impression without reflection]" (figure 7.4). Mendel knew Darwin's hypothesis of pangenesis was incorrect.

In explaining his hypothesis of pangenesis, Darwin referred to "several spermatozoa or pollen-grains being necessary for fertilisation."[37] This supposition directly contradicted Mendel's deduction that one pollen grain unites with one germ cell at fertilization, resulting in parental equality. Although his experiments had already discredited this notion, Mendel still made the effort to directly test Darwin's claim. He meticulously isolated individual pollen grains under a microscope and placed just one pollen grain on each of several different flowers that had their pollen-producing organs removed. The fact that most of these flowers produced a seed con-

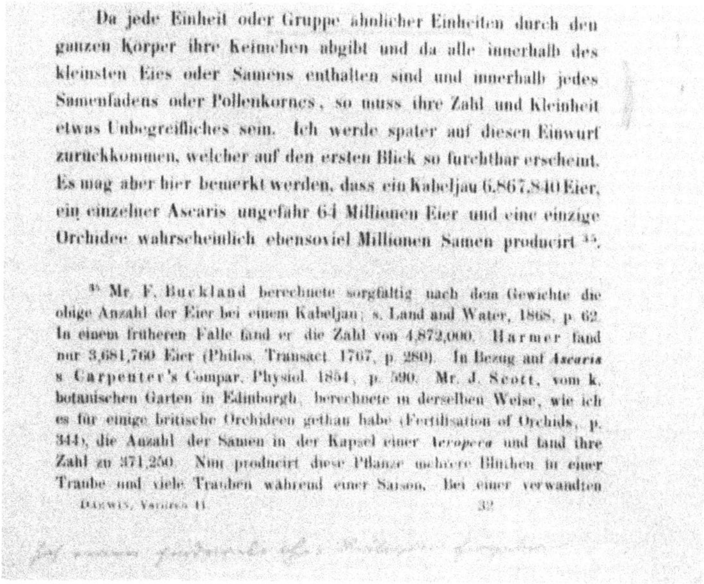

Da jede Einheit oder Gruppe ähnlicher Einheiten durch den ganzen Körper ihre Keimchen abgibt und da alle innerhalb des kleinsten Eies oder Samens enthalten sind und innerhalb jedes Samenfadens oder Pollenkornes, so muss ihre Zahl und Kleinheit etwas Unbegreifliches sein. Ich werde später auf diesen Einwurf zurückkommen, welcher auf den ersten Blick so furchtbar erscheint. Es mag aber hier bemerkt werden, dass ein Kabeljau 6,867,840 Eier, ein einzelner Ascaris ungefähr 64 Millionen Eier und eine einzige Orchidee wahrscheinlich ebensoviel Millionen Samen producirt [35].

[35] Mr. F. Buckland berechnete sorgfältig nach dem Gewichte die obige Anzahl der Eier bei einem Kabeljau; s. Land and Water, 1868, p. 62. In einem früheren Falle fand er die Zahl von 4,872,000. Harmer fand nur 3,681,760 Eier (Philos. Transact. 1767, p. 280). In Bezug auf *Ascaris* s. Carpenter's Compar. Physiol. 1854, p. 590. Mr. J. Scott, vom k. botanischen Garten in Edinburgh, berechnete in derselben Weise, wie ich es für einige britische Orchideen gethan habe (Fertilisation of Orchids, p. 344), die Anzahl der Samen in der Kapsel einer *Acropera* und fand ihre Zahl zu 371,250. Nun producirt diese Pflanze mehrere Blüthen in einer Traube und viele Trauben während einer Saison. Bei einer verwandten

Darwin, Variiren II. 32

Figure 7.4. A few of Mendel's numerous annotations in Darwin's chapter "The Provisional Hypothesis of Pangenesis." Mendel's handwritten note at the bottom of the page reads, *"sich einem Eindrucke ohne Reflexion hingeben,"* meaning "to indulge in an impression without reflection." *Photograph by Daniel J. Fairbanks.*

firmed that a single pollen grain *is* sufficient for fertilization, confirming Mendel's earlier deduction that both parents contribute equally to their offspring. Referring to this observation, Mendel wrote in a letter,

> But one experiment seemed to me to be so important that I could not bring myself to postpone it to some later date. It concerns the opinion of Naudin and Darwin that a single pollen grain does not suffice for fertilization of the ovule. I used *Mirabilis jalappa* for an experimental plant, as Naudin had done; the result of my experiment, however, is completely different. From fertilization with single pollen grains, I obtained 18 well developed seeds, and from these an equal number of plants, of which 10 are already in bloom.[38]

Mendel's words in this passage directly allude to a passage he marked in Darwin's chapter on pangenesis, which reads, in Darwin's original English, "even thirty grains did not fertilise a single seed; but when forty grains were applied to the stigma, a few seeds of small size were formed."[39]

Mendel also compared the results of several of his other experiments to Darwin's conclusions and lamented in a letter that "Darwin's statements concerning hybrids of the genera mentioned in 'The Variation of Animals and Plants Under Domestication,' based on reports of others, need to be corrected in many respects."[40] Although these passages point to errors Mendel found in the details of Darwin's work, Mendel praised Darwin when he believed it was deserved:

> Darwin and Virchow have pointed to the high degree of independence that is typical for individual characters and whole groups of characters in animals and plants. The behavior of plant hybrids indisputably furnishes an important proof of the correctness of this point of view.[41]

Mendel also adopted Darwinian language when expressing his own ideas regarding the evolution of new species in nature. In a letter, he speculated that if two existing species were to hybridize, the parental species might eventually suffer extinction because any new species arising from the hybrids might outcompete their original parental species because they are better adapted to prevailing conditions. In his words, "the naturally occurring hybridizations . . . , if they were repeated often or became permanent, would finally result in the disappearances of the species involved, while one or another of the more happily organized progeny, better adapted to the prevailing telluric or cosmic conditions, might take up the struggle for existence successfully and continue it for a long stretch of time, until finally the same fate overtook it."[42] As Sekerák has pointed out, this inference by Mendel is clearly reflective of Darwin, showing Mendel's perception of how hybridization results in the evolution of new species.[43] The phrase translated here as "struggle for existence" is quintessentially Darwinian. Mendel's original German is "*Kampf ums Dasein*," which is the corresponding phrase in his German translations of Darwin's books. Mendel's biographer, Hugo Iltis, after citing this passage, was moved to exclaim, "The clericalists who stigmatized Mendel the liberal as a Darwinist were not so far wrong!"[44]

An important question, often overlooked, is not why Darwin failed to discover Mendel's writings but rather why none of Darwin's many collaborators learned of Mendel and informed Darwin of him. Some historians have claimed that Darwin might have ignored Mendel because his understanding of heredity was incompatible with Mendel's. However, Darwin often collaborated with expert botanists, several of them in German-speaking parts of Europe. They should have recognized the importance of Mendel's theory and its relevance to Darwinism. Unfortunately, available evidence suggests that neither Darwin nor those associated with him knew of Mendel's article—*with two very important exceptions.*

Libraries throughout Europe and England (and even nine libraries in the United States) carried subscriptions to the journal that contains Mendel's article, so it was readily available to researchers albeit buried in a mass of hundreds of similar articles in volumes of various journals. It is no surprise that it failed to stand out in the mass of similar research. Mendel, however, had a means available to him to directly inform a select group of scientists about his research. He obtained forty reprints of his paper from the printer, probably in December 1866. He sent several of them to notable botanists with accompanying letters. Darwin corresponded with two of those botanists, and the histories of the reprints Mendel sent them are well known.

On New Year's Day 1867, Mendel sent a reprint to Anton Kerner von Marilaun, a botanist and plant hybridist at the University of Innsbruck in Austria. Mendel knew Kerner from the time when both were students studying with Unger and Fenzl at the University of Vienna in the early 1850s. Whether Kerner corresponded with Mendel immediately after receiving the reprint is unknown. Some have mistakenly claimed that Kerner had little interest in the reprint because it was supposedly uncut.[45] However, I have personally viewed the reprint Kerner owned. It is cut, and it is the one depicted in figure 7.3 with Mendel's handwritten corrections. Kerner kept the reprint and the accompanying letter, taking them with him when in 1878 he took a position as professor of natural history and curator of the botanical garden at the University of Vienna, where the reprint is currently archived. Kerner did not forget Mendel, referring to him favorably in a letter dated 1872.[46] Kerner corresponded with Darwin, and Darwin cited him in his writings, but there is no evidence that Kerner informed Darwin of Mendel.

The second reprint with a well-known history is the most important one. Mendel sent it to the famous botanist Carl von Nägeli in Munich, with a lengthy accompanying letter dated December 31, 1866. This reprint and the letter initiated scientific collaboration with detailed correspondence between Mendel and Nägeli that lasted until 1873. Nägeli retained Mendel's letters, studying them and writing his own notes on them. These letters were discovered after both Mendel and Nägeli had died, and they are the principal source of information about Mendel's scientific pursuits during the years he wrote them.

At the same time Nägeli was corresponding with Mendel, he was also corresponding with Darwin. Nägeli responded to Mendel's letter on

February 24, 1867, then wrote his first letter to Darwin a little more than a month later, on March 31, 1867. He continued to correspond with both Mendel and Darwin after that time. Although some of Nägeli's letters to Darwin have been lost, there is no indication that he mentioned Mendel in any of them. Exacerbating Nägeli's apparent failure to mention Mendel is the fact that Mendel addressed Darwin's writings in three of his letters to Nägeli.[47]

Nägeli was the most prominent scientist who could have raised awareness of Mendel's research, and he could easily have notified Darwin of it. Although he spent abundant time corresponding and collaborating with Mendel for seven years, he failed to take much interest in Mendel's pea experiments. Instead, most of Nägeli's correspondence with Mendel and with Darwin addressed a plant known as hawkweed, which he, Mendel, and Darwin consistently referred to by its Latinized scientific name *Hieracium*. It is regrettable for Nägeli's legacy that despite the many important contributions he made to science, he will forever be remembered as the person who could have lifted Mendel's groundbreaking discovery from obscurity to prominence and connected Mendel with Darwin but failed to do either.

The Steadfast Scientist 8

the hybridization experiments of which I have become so fond

—MENDEL IN A LETTER TO NÄGELI

MENDEL'S MOST IMPORTANT WORK was *Experiments on Plant Hybrids*, the most extensive and detailed article he wrote in his lifetime. Although it posthumously elevated him from obscurity to fame, it was only a part of his long and productive scientific career. As mentioned in chapter 4, Mendel had already published two research articles in the *Proceedings of the Imperial-Royal Zoological-Botanical Society in Vienna* before publishing his famous 1866 article. Like many scientists of his day, he did not specialize in a single discipline. He had an intense interest in cells as the basis of life, a topic he studied under Unger, and he owned three microscopes, the latest obtained in the 1870s. He maintained an extensive set of cellular preparations preserved in glass microscope slides, most consisting of plant materials but a few of animal and human specimens, which he probably obtained from vendors rather than preparing them himself.[1] He devoted much time to meteorology and beekeeping alongside his plant hybridization research. He would present one more lecture to the Natural Science Society on plant hybrids in 1869, publishing it in the society's proceedings in 1870. This would be his last publication on plant hybrids but not the end of his hybridization experiments, which he continued until 1873. Unfortunately, most of his remarkable discoveries never reached the public during his lifetime, and much of the detailed information on them is lost forever. Referring to Mendel's post-1866 years, the famous twentieth-century geneticist Alfred Sturtevant lamented,

> The picture that emerges is of a man very actively and effectively experimenting, aware of the importance of his discovery, and testing and extending it on a wide variety of forms. None of these results were published; it is difficult to suppose that his work would have been so completely ignored if he had presented this confirmatory evidence.[2]

Why did Mendel fail to publish what was obviously an abundant record of valuable scientific data, evidence, and interpretation? The most likely answer comes from his nephew Ferdinand Schindler, son of his sister Theresia. Ferdinand and his brother Alois lived in Brünn during the latter years of Mendel's life while both were studying medicine. The two spent considerable time with their uncle, discussing science and playing chess. In 1905, after Mendel had achieved worldwide fame, Ferdinand wrote a letter reminiscing on the conversations he and his brother had with him. Part of that letter offers evidence that Mendel had manuscripts written and ready for publication, but he felt constrained. The grammatical errors in this excerpt from Schindler's letter are in the original letter; his native language was German, but he wrote this letter in English:

> The died abbot Mendel was a man of liberal principles. . . . He readed with the greatest interest Darwin's works in the German translation and admired his genius, though he did not agree to all principles of this immortal natural philosopher. But it can be, that my uncle in the latter part of his life, retired from scientical evolutionary questions, because he had many clerical enemies. He said often to us nephews, that we shall find at his heritage, papers for publication, that he could not publish during his life. But we did not receive anything from the cloister, not even a thing for remembrance.[3]

There is evidence confirming Schindler's memory of unpublished manuscripts and their tragic fate. Antonín Doupovec, whose mother cared for Mendel during his final days, often accompanied her in her visits and grew close to Mendel. He recalled that "in the wardrobe of his room, thousands of sheets of paper covered with scientific notes and data were found after his death. I myself had them in my hands."[4] And one of the young friars whom Mendel recruited, Pater Clemens Janetschek, who grew close to his abbot, had a similar recollection. According to him, Mendel left a treasure trove of papers when he died that would have been priceless had they survived, but, according to Janetschek, they were "committed to the flames."[5] Documents that Ferdinand Schindler was able to preserve, such as letters and items from outside the monastery, were destroyed during World War II.[6]

Mendel was elected abbot in 1868 while he was amid some of his most extensive research on plant hybrids. Until then, Abbot Napp had provided a layer of protection between Mendel and Bishop Schaffgotsch, who had previously condemned Mendel and his monastic colleagues for secularism. Now, Mendel's election as abbot placed him in a much more precarious position, having to work directly with the bishop and other church authorities. His abbatial years, as we will see in the upcoming chapter, became increasingly tormented, as his relationships with his "clerical enemies" (as his nephew put it) grew strained and ultimately intolerable. Mendel perceived his research as consistent with the writings of Darwin and Unger, which put him at odds with a strong antievolutionary movement among the Catholic clergy, centered in Vienna but also powerful in Brünn.[7] If Ferdinand Schindler was correct, Mendel felt it best to leave his manuscripts to his nephews for posthumous publication. To the detriment of science, those manuscripts were apparently burned, never reaching his nephews.

Much of what we do know of Mendel's work after his famous 1866 article is found in a series of lengthy and informative letters, alluded to in the previous chapter, that he wrote to the famous botanist and evolutionist Carl von Nägeli (figure 8.1). At the time, Nägeli was professor and director of the Botanical Garden in Munich and well known for his work in botany and plant physiology. Were it not for Mendel, Nägeli's name might have faded in the history of science, recalled by only a few avid devotees of nineteenth-century German botanical history. Nägeli's place is now firmly cemented in history *because* of his association with Mendel.

Figure 8.1. Carl von Nägeli, professor and director of the botanical garden in Munich, as he appeared during the time he corresponded with Mendel. *Drawing by Daniel J. Fairbanks based on a historic photograph, sepia pastel on paper. Collection of the artist.*

Some authors claim that Nägeli disrespected Mendel while leading him down a disastrous research path. For example, Siddhartha Mukherjee, in his excellent award-winning book *The Gene: An Intimate History*, assumed that Nägeli, after replying to several of Mendel's letters, "could hardly be bothered with the progressively lunatic ramblings of a self-taught monk in Brno."[8] According to Robin Marantz Henig in her book *The Monk in the Garden*, "Because of misdirection, either accidental or deliberate, from the great Nägeli, Mendel lost faith in his own results."[9] The texts of Mendel's letters to Nägeli, however, reveal a much more respectful, elaborate, and scientifically informative collaboration between the two scientists than these characterizations suggest.

Mendel wrote his first letter to Nägeli on New Year's Eve 1866, enclosing a reprint of his paper. Fortunately, Nägeli kept Mendel's letters, although one was lost in the mail, and there is evidence of a missing page in another.[10] Nägeli's heirs fortunately spared the letters from destruction, and his student Carl Correns found them long after both Mendel and Nägeli had died.

Correns published the letters in 1905, after Mendel had achieved posthumous fame. A fragment of the first letter Nägeli sent to Mendel somehow escaped the flames at the monastery, and Pater Clemens Janetschek gave it to Mendel's biographer, Hugo Iltis, who published its contents in his biography of Mendel.[11] The fact that most of Nägeli's letters to Mendel were lost or destroyed means that we must infer what Nägeli wrote from Mendel's responses and from the marginal notes Nägeli wrote on the letters he received from Mendel.

In Mendel's first letter to Nägeli, he briefly referred to the enclosed reprint on his pea research, then lamented,

> The results which Gärtner obtained in his experiments are known to me; I have repeated his work and have re-examined it carefully to find, if possible, an agreement with those laws of development which I found to be true for my experimental plant [peas]. However, try as I would, I was unable to follow his experiments completely, not in a single case! It is very regrettable that this worthy man did not publish a detailed description of his individual experiments, and that he did not diagnose his hybrid types sufficiently. . . . A decision can be reached only when new experiments in which the degree of kinship between hybrid forms and their parental species are precisely determined, rather than simply estimated from general impressions, are performed.[12]

In this first letter, Mendel informed Nägeli that he had started experiments with species other than the garden pea and the common bean to confirm whether the patterns he observed in these two species were generalizable. As he put it, "If, for two differing traits, the same ratios and series which exist in *Pisum* can be found, the whole matter would be decided."[13] His approach, however, was somewhat unusual, even daring. Among the several species he chose to research, he intentionally selected two that he knew were likely to contradict his observations in the garden pea. In other words, he sought out the exceptions to obtain a broader understanding of the general nature of inheritance in plants. He spent a paragraph in the letter describing the experiments he had started with the ornamental flower known as avens, its Latinized scientific name *Geum*. Mendel noted, "This plant, according to Gärtner, belongs to the few known hybrids which produce non-variable progeny as they remain self-pollinated." Mendel, however, did not fully accept Gärtner's opinion, surmising that there was a "definite law" that governed *Geum* hybrids and that, "if it could be discovered, would also give clues to the behavior of other hybrids of this type."[14]

Mendel then turned his attention in the letter to the hawkweeds, their Latin name *Hieracium*, a group of wildflowers with numerous related species. "The surmise that some species of *Hieracium*, if hybridized, would behave in a similar fashion to *Geum*, is perhaps not without foundation," he wrote, presuming that their patterns of inheritance might display at least some degree of constancy.[15] Botanists across central Europe had chosen to study *Hieracium*, including at least three of Mendel's botanical colleagues in Brünn, and Nägeli was the leading expert on this group of species.

As the concluding sentence of this first letter, Mendel wrote, "I am afraid that in the course of my experiments, especially with *Hieracium*, I shall encounter many difficulties, and therefore I am turning confidently to your honor with the request that you not deny me your esteemed interest when I need your advice."[16] From that point forward, Mendel and Nägeli carried on detailed correspondence, mostly about *Hieracium*, for six and a half years.

Nägeli responded to Mendel's letter, and, fortunately, this first letter is the one that escaped the flames at Mendel's death. In it, Nägeli briefly addressed the pea experiments:

> It seems to me that the experiments with *Pisum*, far from being finished, are only beginning. The mistake made by all the more recent experimenters is that they have shown so much less perseverance than Kölreuter and Gärtner. I note with pleasure that you are not making this mistake, and that you are treading in the footsteps of your famous predecessors. You

> should, however, try to excel them, and in my view this will only be possible (and thus alone can any advance be made in hybridization) if experiments of an exhaustive character are made upon one single object in every conceivable direction. No such complete set of experiments, providing irrefutable proofs for the most momentous conclusions, has ever been made. If as a result of your hybrid fertilizations, you have a store of seeds which you are not yourself going to plant, I should be very glad to cultivate them in our own garden here in order to see whether they remain constant in different surroundings.[17]

Nägeli then went into some detail describing the types of hybrid seeds and their progeny that he wished Mendel to send, if available, specifying them by Mendel's designations using various combinations of the letters *A*, *a*, *B*, and *b*. Nägeli also enclosed reprints of several of his own articles.

Some have interpreted Nägeli's tone as condescending, a claim that has merit in that Nägeli somehow did not sufficiently recognize the brilliance and exhaustive nature of Mendel's experimental approach. Even so, Nägeli praised Mendel for not making the mistakes of "all the more recent experimenters" and commending him for "treading in the footsteps of your famous predecessors." He legitimately questioned whether the results Mendel observed would be evident in another environment, implying that a different environment might affect the outcome. To his credit, Nägeli offered to assist Mendel in experimenting on this topic, requesting that Mendel send him pea seeds from his experiments, specifying the exact types he wished Mendel to send.

Nägeli then turned his attention in the letter to *Hieracium*:

> Your design to experiment on other plants of other kinds is excellent, and I am convinced that with these different forms you will get notably different results (in respect of the inherited characters). It would seem to me especially valuable if you were able to effect hybrid fertilizations in *Hieracium*, for this will soon be the species about whose intermediate forms we shall have the most precise knowledge.[18]

Despite Nägeli's criticism, Mendel was elated to receive a response from one of the world's most famous botanists. His enthusiasm is evident in the first few sentences of his reply: "My most cordial thanks for the printed matter you have so kindly sent me! . . . This thorough revision of the theory of hybrids according to contemporary science was most welcome. Thank you again!"[19]

Mendel's second letter to Nägeli is the longest of the letters he wrote to him. Among the many details evident in this letter is one that clarifies much of the historical record on how and when Mendel conducted his pea experiments. He specified the years as 1856 to 1863 and then spoke of the need to verify his results in other species. His presentation in 1865 was, as he explained, an attempt to inspire other researchers to experimentally verify his results and his conclusions. He then stated, the melancholy palpable in his words, "as far as I know, no one undertook to repeat the experiments."[20]

Next, he responded to Nägeli's skepticism: "I am not surprised to hear your honor speak of my experiments with mistrustful caution; I would not do otherwise in a similar case."[21] Mendel was apparently also elated at Nägeli's offer to plant pea seeds from Mendel's experiments in his garden in Munich to "see whether they remain constant in different surroundings." Mendel prepared a series of carefully numbered and cataloged packets of pea seeds for Nägeli to cultivate, anticipating that he would observe firsthand the same numerical patterns. The bulk of this letter contains detailed information about the seeds in these packets.

The letter, with its accompanying seed packets, was dated April 18, 1867, just in time for planting. Despite Mendel's meticulous efforts to prepare and catalog these seeds, there is no information to indicate whether Nägeli made the effort to plant them; none of Mendel's subsequent letters to Nägeli mention the pea seeds he sent. Had Nägeli done so, he would have observed firsthand the ratios that Mendel described.

By this time, Nägeli had developed his own (incorrect) hypothesis of inheritance that differed from Mendel's theory, and this was probably one of several reasons for his skepticism. The remaining eight letters Mendel would send to Nägeli through 1873 refer mostly to *Hieracium* and some of Mendel's experiments with other species. Only on rare occasions did Mendel refer to his pea experiments in these subsequent letters. Nonetheless, these rare references ended up being some of the most informative passages in his letters.

In this second letter, Mendel listed in detail the various species he had chosen to research to determine if their inheritance behaved the same as in peas, using their Latin scientific names, not their common names, although I will list only their common names here: columbine, stocks, four o'clock, bellflower, plume thistle, avens, toadflax, mullein, nasturtium, lady's purse, morning glory, wallflower, snapdragon, maize (corn), and hawkweed. Mendel would, however, devote most of his attention to hawkweed (*Hieracium*), a type of wildflower that is especially common and varied in

Europe with many named species. It typically has bright yellow flowers (or, more rarely, orange or red) that look like dandelion flowers. In fact, it is closely related to dandelions, sunflowers, asters, and daisies.

Historians have long blamed Nägeli, claiming that he led Mendel along a disastrous path by encouraging him to study *Hieracium*. According to a famed geneticist named Alfred Sturtevant, the *Hieracium* research "must have been a great disappointment to Mendel," and *Hieracium* "was the worst possible choice."[22] Henig called it "a completely misguided choice of plant for Mendel to work on."[23] Mukherjee lamented that it was "a catastrophically wrong choice" and that "the results were a mess."[24] Iltis concluded that the results of these experiments "remained an enigma" and that they "shattered the hopes he had entertained of finding a confirmation of the principles of inheritance worked out by him in the case of *Pisum*, and thus establishing these principles as universally valid laws."[25]

By contrast, in 1996, Orel drew the opposite conclusion, stating that, "Mendel's choice of *Hieracium* as a new experimental plant was, however, in no way 'unfortunate.'"[26] In fact, after almost seven years of corresponding with Mendel, Nägeli had grown to admire him, praising him by name in his massive book on *Hieracium* the year after Mendel died.[27]

Although Nägeli is often blamed in hindsight for encouraging Mendel to research *Hieracium*, it was Mendel who enthusiastically approached Nägeli, asking for advice on *Hieracium*. Nägeli then confirmed that Mendel had made the right decision in choosing to research *Hieracium*. So why did Mendel choose *Hieracium*, and why have some authors considered it in retrospect to have been such a disastrous choice?

At the time Mendel began his experiments on *Hieracium*, the diverse group of species that fell under this classification was considered as an ideal model for research on plant evolution. Darwin relied on *Hieracium* as an example of natural selection in *Origin of Species*, and Nägeli had published much on it, corresponding with Darwin about it.

Closer to home, several of Mendel's friends and collaborators were already studying *Hieracium*. Makowsky published in 1862 a compilation of plant species in the surrounding region, noting natural hybrids between different species of *Hieracium*. The following year, Makowsky presented to the society his observations of natural *Hieracium* hybrids in southern Moravia. Then, later that year, Niessl, also a founding member of the Natural Science Society, reported his observations of natural hybrids between different *Hieracium* species in northwestern Moravia. At the December 14, 1864, meeting, which was just two months before Mendel would present his pea research, Niessl again reported on *Hieracium* hybrids observed in nature in southern Moravia.[28] In the January 1865 meeting, the meeting that immediately preceded Mendel's presentation on his pea research, Makowsky quoted Darwin's reference to *Hieracium* as an example of a type of plant for which "hardly two naturalists can agree which forms to rank as species and which as varieties."[29]

In choosing the pea plant for his experiments, Mendel had focused on a domesticated species, confined to vegetable gardens, entirely dependent on humans and incapable of persisting in the wild. Darwin and others had claimed that variation in domesticated species differed from that of wild species. Mendel hoped to discover the role of hybridization in the evolution of new species in nature, so he needed to turn his attention away from domesticated species to plants growing wild. It is no surprise that he wrote to Nägeli that he had chosen to focus on *Hieracium*. Mendel's own words, in a lecture he gave to the Natural Science Society in 1869 and published in 1870, make it clear why he believed *Hieracium* to be the optimal choice:

> To indicate the object with which these fertilization experiments were undertaken, I venture to make some preliminary remarks respecting the genus *Hieracium*. This genus possesses such an extraordinary profusion of distinct forms that no other genus of plants can compare with it. . . . Regarding no other genus has so much been written or have so many and such fierce controversies arisen, without as yet coming to a definite conclusion. . . . Regarding the question whether and to what extent hybridization plays a part in the production of this wealth of forms, we find various and conflicting views held by leading botanists. . . . The question of the origin of the numerous and constant intermediate forms has recently acquired no small interest since a famous *Hieracium* specialist has, in the spirit of the Darwinian teaching, defended the view that these forms are to be regarded as arising from the transmutation of lost or still existing species.[30]

Mendel's reference to "the Darwinian teaching" is his first and only mention of Darwin in his printed publications, though he would later mention Darwin three times in his letters to Nägeli. And the "the famous *Hieracium* specialist" mentioned in this passage undoubtedly is Nägeli.

In the third letter, Mendel lightheartedly admits being unable to sufficiently study *Hieracium* in its natural environment: "I am no longer very capable of botanical excursions, because heaven has blessed me with excess weight, which, as a result of general gravitation, becomes very perceptible on extended walks, especially when climbing."[31] Mendel had taken up cigar smoking because he believed it to be a remedy for obesity.[32]

Some historians take Nägeli's skepticism as evidence that he was dismissive and uninterested in Mendel's experiments. On the contrary, the letters from 1867 through 1873 portray extensive collaboration and mutual respect between Nägeli and Mendel, especially as related to *Hieracium* hybridization. In several instances, Mendel mentions seeds and live plants sent to him by Nägeli and seeds and plants Mendel had sent in return. Although their correspondence was extensive, there is no indication that the two ever met in person.

Researching *Hieracium* turned out to be an enormous challenge for both Mendel and Nägeli. Obtaining hybrids was notoriously difficult, although it is obvious from Mendel's letters that, after considerable labor, he was quite successful in obtaining true hybrids. The experiments were so difficult that Mendel nearly had to abandon them. In his eighth letter to Nägeli, penned on July 3, 1870, he wrote,

> I found myself in serious danger of having to renounce my hybridization experiments completely, and this due to my own carelessness. Since diffuse daylight was not adequate for my work on the small *Hieracium* flowers, I had recourse to an illumination apparatus (mirror with convex lens), without suspecting what damage I might have done with it. . . . [A] peculiar fatigue and exhaustion of the eyes appeared and reached a serious degree in spite of my immediately sparing my eyes as much as possible. It made me incapable of any exertion well into the winter.[33]

Despite the difficulties both Mendel and Nägeli faced researching *Hieracium*, they both believed it to be an excellent plant for experimentation because it portrayed so many variations in nature of the sort that lent themselves well to experiments on inheritance.

Neither Nägeli nor Mendel knew at the time that *Hieracium* often reproduces through a process known as apomixis (though not exclusively), wherein the seeds inherit their genetic material only from the female parent and thus are genetically identical to that parent regardless of which male parent contributes pollen. The disparagement of Mendel's and Nägeli's choice to research *Hieracium* is based on the claim that Mendel stumbled on a rare apomictic species that frustrated him to no end because it failed to behave like the garden pea. Apomixis, however, is not rare in plants. One need look no further than a common lawn to find it in abundance. Most lawn grasses, as well as dandelions growing as weeds in a lawn, reproduce through apomixis. It is most common in certain families of plants, especially the daisy family, a very large and diverse plant family that includes *Hieracium*.

Mendel's choice of *Hieracium* was, in fact, well informed and quite intentional. As mentioned in the previous chapter, he concluded his famous article by making a strong point that plant hybrids are of two types: those that produce *variable* offspring and those that produce *constant* offspring. He stated that the pea plant is clearly of the variable type and that certain laws (which he discovered) govern the inheritance of these variations. However, he also discussed the work of previous plant hybridists who had studied plants whose hybrids produced constant offspring.

Mendel was not sure whether *Hieracium* hybrids would display variable or constant offspring, although he expected some degree of constancy and was anxious to find out. He wrote to Nägeli in 1867 that the offspring of his *Hieracium* hybrids "had rooted well, and should flower next year." He then made it clear that he expected them to either be constant or variable: "Whether they will retain the characteristics of the hybrid, or whether they will show variations, will be determined by next year's observations."[34] As Peter van Dijk and T. H. Noel Ellis pointed out, "These are not the words of a frustrated man."[35]

In 1870, Mendel published his results on *Hieracium* hybridization in the same journal where he had published his pea and bean experiments three years earlier. He succinctly pointed out the sharp contrast between the mechanisms of inheritance in the garden pea and *Hieracium*:

> In *Pisum* the hybrids, obtained from the immediate crossing of two forms, all have the same type, but their posterity, on the contrary, are variable and follow a definite law in their variations. In *Hieracium* according to the present experiments the exactly opposite phenomenon seems to be exhibited.[36]

He drew the conclusion (correctly as it turned out) that inheritance was variable in the offspring of pea hybrids and constant in the offspring of *Hieracium* hybrids.

This article ended Mendel's publications on plant hybrids but not his experiments. In 1870, he described in a letter to Nägeli the results from experiments with stocks, four o'clock, and maize (corn), stating that "their hybrids behave exactly like those of *Pisum*."[37] Unfortunately, he never published this confirmation, which turned out to be a significant discovery. Had he done so, it might have solidified his theory and attracted attention to it.

This was not the only set of experimental data Mendel collected yet did not publish. In his eighth, ninth, and tenth letters to Nägeli, all written after the 1870 *Hieracium* article, Mendel described additional large-scale experiments in *Hieracium* and other plants. He never published the results of these later experiments despite how extensive and conclusive they were. These final three letters are important because they contain novel scientific discoveries, preserved only in the texts of the letters. In one of the most important of these experiments, Mendel made a significant discovery that contradicted Darwin—that a single pollen grain is sufficient for fertilization—as detailed in the previous chapter.

During the years from 1868 to 1871, Mendel focused his attention on *Hieracium* for nearly all his hybridization experiments. By 1871, he ended these experiments and sent all 235 of his herbarium specimens to Nägeli.[38] These were not forgotten. In 1885, the year after Mendel's death, Nägeli and Albert Peter published a compilation of *Hieracium* specimens and their botanical descriptions in their book *The Hieracia of Middle-Europe* (*Die Hieracien Mittel-Europas*) and in it referenced Mendel eight times.[39] Mendel's *Hieracium* specimens sent to Nägeli were eventually distributed throughout Europe and currently reside in twenty-two museum herbarium collections.[40]

There is a gap of more than three years between Mendel's ninth letter to Nägeli, sent in 1870, and the final tenth letter in 1873. One reason for this gap is Nägeli's delayed response to the ninth letter, for which Nägeli asked forgiveness because of difficulties caused by an ongoing war. Mendel also referred to a letter he sent that Nägeli never received.

Mendel's final letter to Nägeli contains a note of exasperation: "I am really unhappy about having to neglect my plants and my bees so completely."[41] By this time, Mendel was well into his fifth year as abbot, and administrative duties were taking an enormous toll. In a letter to Nägeli, written more than five years earlier, on May 4, 1868, Mendel stated,

> Recently there has been a completely unexpected turn in my affairs. On March 30 my unimportant self was elected life-long head, by the chapter of the monastery to which I belong. From the very modest position of teacher of experimental physics I thus find myself moved into a sphere in which much appears strange to me, and it will take some time and effort before I feel at home in it. This shall not prevent me from continuing the hybridization experiments of which I have become so fond; I even hope to be able to devote more time and attention to them, once I have become familiar with my new position.[42]

For the first five years of his abbacy, Mendel was true to his word, conducting extensive hybridization experiments and publishing his *Hieracium* article. The six letters Mendel wrote to Nägeli during these early years of his abbacy portray a dedicated researcher spending enormous time on meticulous and large-scale hybridization experiments, cultivating thousands of plants, and exchanging seeds and live plants with Nägeli. Regrettably, much of the abundant data Mendel collected during these years, along with novel and important scientific discoveries, remained unpublished, their contribution to science confined to letters between two collaborating scientists discovered years after both had died.

Mendel did not restrict his scientific research to plant hybridization. Much of his training was in physics, and he applied its principles to meteorology. He regularly measured wind speed and direction, barometric pressure, temperature, rainfall, groundwater levels, and atmospheric ozone concentrations with instruments he set up at the monastery. He recorded sunspots and their movements, carefully drawing their changing positions on circular maps of the sun. He dutifully reported annual meteorological data that were published in the same journal where he published his classic article.[43]

On October 13, 1870, Mendel observed a tornado as it struck the monastery head-on. It was, as he described it, "a truly infernal symphony accompanied by the shattering of the window panes, the demonic rumbling of roof tiles, and slates hurled through the smashed windows, some as far as the opposite walls of the room."[44] In the midst of this mayhem, Mendel had the presence of mind to notice that the tornado was rotating in the wrong direction. In the northern hemisphere, tornadoes rotate in counterclockwise direction, but Mendel noticed the clockwise direction of this one. To scientifically confirm this unusual observation, he measured the angles at which the double-pane glass in the monastery windows had broken, reconstructing from these angles the curving trajectory of projectiles that had entered through the broken windows and landed in the rooms of the monastery and church. He then published his observations in an 1871 article in the same journal that carried his earlier articles on plant hybridization.[45] In it, he proposed a hypothetical explanation of the unusual rotation but cautioned against placing too much confidence in his hypothesis with a clever play on words:

> And with that the discussion of our dangerous guest of the 13th of October draws to a close. We have exhausted ourselves with various conjectures about it but must finally admit that even with our best intentions we could proceed no further than an airy hypothesis, constructed with airy materials, and built on a most airy foundation.[46]

About this time, weather forecasting had become a major focus of meteorological research, especially given its benefits for agriculture and safety, and it captured Mendel's interest. Mendel published an article on this topic, "The Basis of Weather Forecasting (*Die Grundlage der Wetterprognosen*)," in 1879, six years after he had abandoned his plant hybridization experiments.[47] And he continued to publish additional information on his meteorological observations.[48]

Mendel's interest in beekeeping began about 1870. That year, he attended a large professional conference on apiculture, visited accomplished bee breeders, and had an elaborate brick structure constructed on a hill overlooking the monastery to accommodate a collection of honeybee hives.[49] He built innovative mating cages to ensure that queen bees mated only with the drones he chose, and he imported different types of bees for breeding from various parts of Europe as well as from Cyprus and Egypt. He regularly attended the monthly meetings of the local apicultural society, presenting results of his experiments and techniques to fellow beekeepers. Some historians have suggested that he conducted hybridization experiments in bees to confirm his theory of heredity in the animal kingdom.[50] It appears, however, that he simply may have intended to breed better honey-producing strains for practical purposes. He often shared the results of his research in meetings, and those results consist of a wide range of pragmatic aspects of beekeeping, including hive maintenance, disease infestations, and the best plants to cultivate near the hives for the sweetest honey. A newspaper article in May 1873 announced that Mendel was overseeing, at the monastery's expense, revegetation of the Spielberg Hill with flowering plants for bees to visit.[51]

The information about Mendel's work on beekeeping and breeding comes mostly from the recorded minutes of meetings where he gave presentations and contributed to discussions. He was very active in collaborating with other beekeepers, both local and distant. Some information about his research is from articles published by other beekeepers, one referring to Mendel as "an eminent expert" and another as "one of the most experienced beekeepers in Moravia" and as the "outstanding expert Mendel."[52]

He was also well respected as an agriculturalist and plant breeder, with a focus on ornamental flowers and fruit trees. In an 1862 photograph, some members of St. Thomas Monastery hold props in their hands to portray their interests. The preferred props are books, but Mendel holds a fuchsia flower, setting him apart from the others (see figure 2.2). Years later, in 1882, a plant breeder named a new fuchsia variety "Prelate Mendel" in honor of Mendel's contributions to fuchsia breeding. Mendel also took considerable interest in fruit breeding, especially grapes, apricots, apples, and pears, receiving medals for his new varieties of apples and pears.[53]

Although Mendel's election as abbot did not deter him from continuing his experiments on plant hybrids, his abbatial and other administrative duties would progressively demand more of his time. To his benefit, his position as abbot and prelate was a prestigious one, leading to other appointments of high responsibility, such as president of the Moravian Mortgage Bank and vice president of the Natural Science Society in Brünn.[54] Lamentably, his interactions with church superiors and government officials increasingly tormented him to the point that he would, in his final years, become obsessively, even pathologically, suspicious and distrustful.

9 The Resolute Abbot

Gracious, generous, and affable to everyone,
Fraternal father to us brethren was he,
Flowers he loved and he fought to defend the law.

—PATER CLEMENS D'ELPIDIO JANETSCHEK,
A YOUNG FRIAR'S ELEGY TO HIS ABBOT

CYRILL FRANZ NAPP, MITERED PRELATE, distinguished scholar, and abbot of the Order of St. Thomas in Altbrünn, died at the age of seventy-two on Mendel's forty-fifth birthday, July 22, 1867. His abbacy had encompassed a remarkable forty-three years. For twenty-five of those years, he had employed his immense benevolence and political prowess to enable Mendel's career as a scientist and teacher in the face of oppressive opposition from his ecclesiastical superiors. Mendel's admiration for Abbot Napp was boundless. It was, therefore, a gallant tribute when, seven months after Napp's passing, the community of friars elected Mendel to succeed Napp as abbot and prelate for the remainder of his life (figure 9.1).

The election was not initially unanimous. The first two ballots were indecisive with no clear majority for any candidate. However, Bratránek voluntarily stepped aside as a candidate, preferring to continue his prestigious university position at the Jagiellonian University in Krakow and lending his support to Mendel. In the third round, Mendel received a decisive twelve votes. He did not vote for himself; in all three ballots, Mendel dedicated his vote to Klácel.[1]

Figure 9.1. Painting of Mendel as Abbot and Prelate, known as the "Great Prelate Portrait." *Historic photograph (Iltis, 1924), public domain.*

The newspaper *Daily Messenger of Moravia and Silesia* (*Tagesbote aus Mähren und Schlesien*) proclaimed on April 1, 1868,

> The population greets the election with undivided joy. We are informed by many of the citizens of Altbrünn that a proposal is afoot to deliver a congratulatory address to the prelate. This time, at any rate, there is justification for the Latin adage: "Vox populi, vox dei" [the voice of the people is the voice of God].[2]

Mendel wrote to Nägeli,

> Recently there has been a completely unexpected turn in my affairs. On March 30, my unimportant self was elected life-long head, by the chapter of the monastery to which I belong. From the very modest position of teacher of experimental physics I thus find myself moved into a sphere in which much appears strange to me, and it will take some time and effort before I feel at home in it.[3]

Mendel's abbacy lasted almost sixteen years. It was a time of prestige, honor, and positions of influence. He continued to pursue his interests in *Hieracium* research, flower and fruit breeding, beekeeping, and meteorology. He occasionally visited his home village, where he founded a fire brigade and financed construction of a firehouse just a few steps from his childhood home. The fire station still stands and bears his name. He also spent much time with his nephews, Johann, Alois, and Ferdinand Schindler, who were his sister Theresia's sons, and lived in Brünn. He was generous with them, supporting them financially.

Johann Schindler was the oldest. Like his uncle, he became an accomplished scientist and teacher. Tragically, he died prematurely in his mid-twenties while serving as a science teacher in Brünn. He was a member of the Natural Science Society in Brünn, attending the monthly meetings with his uncle. An announcement in the society's journal, the same journal in which Mendel published his research, states that in 1881, Johann Schindler "was torn from us through death."[4] The cause was tuberculosis, which had afflicted many others at the time. His brothers, Alois and Ferdinand, studied medicine and became prominent physicians, living to see their uncle achieve posthumous scientific renown.

As abbot and prelate, Mendel was expected to deliver sermons. Yet precious little remains to inform us of his preaching. His sermons were not published, nor is there any known commentary from others revealing anything about them. Nonetheless, notes for two Easter sermons dating to the time of his abbacy were found among his notes on a dispute regarding taxation of the monastery. They are not complete sermon texts, suggesting that Mendel may have preached in his own words from notes rather than reading from a text. In one, he begins with the statement of the risen Christ: "Blessed are those who do not see and yet believe. Truly blessed" ("*Selig, die nicht sehen und doch glauben. Wahrhaft selig!*"). He then refers to the risen Christ appearing to St. Mary Magdalene (the saint of his birthday) as a gardener as his way of introducing the imagery of God as a gardener planting the seed of grace within a person's soul, which then grows into a plant through nourishment from soil, rain, and warmth from the sun. In this manner, according to Mendel, the natural and the supernatural unite, and good works are sanctified by the grace of God. The metaphorical imagery of the gardener, seed, soil, rain, and sunshine reflect Mendel's passion for plants in his religious teachings.[5]

Mendel held several prestigious positions, among them membership on the Central Committee of the Moravian-Silesian Agricultural Society, where he was very active and consistently attended society meetings until shortly before his death. He was elected vice president of the Natural Science Society in Brünn, which he had helped to found. He also served as vice president and, for two years, president of the Moravian Mortgage Bank, which provided him with a lucrative salary.[6]

He was widely known as a supporter of the arts. Mendel commissioned ceiling paintings in the prelacy, also known as the great chapter hall, depicting his interests in fruit tree grafting, meteorology, beekeeping, plowing the soil, and religious worship.[7] Unfortunately, these paintings no longer exist, having been removed during twentieth-century renovations. On the floor immediately aboveground is an ornate library. Mendel also commissioned ceiling paintings for the library, and these remain intact. One depicts his abbatial emblem, and others portray arrangements of the plants that most interested him, including the garden pea with tendrils among ornamental flowers (figure 9.2).[8]

According to Iltis, Mendel's closest friend among his fellow friars during his abbacy was Pavel Karel Křížkovský.[9] As mentioned in earlier chapters, Křížkovský was a musical composer, choirmaster, teacher, and fellow student with Mendel in Troppau and Olmütz during their teenage years before either had entered the monastery. In 1865, the same year Mendel presented his research, eleven-year-old Leoš Janáček began boarding at the monastery where he studied music under Křížkovský.[10] Janáček eventually rose to fame as one of the most illustrious and beloved of Czech composers. When Mendel died, Janáček played the organ and conducted the requiem at the funeral. While serving as abbot in 1872, Mendel received

Figure 9.2. Two of the ceiling paintings in the monastery library commissioned by Mendel. The upper one is Mendel's abbatial emblem, and the lower one is one of several paintings with plants, including ornamental flowers and the plants Mendel researched, among them the garden pea. *Photographs by Daniel J. Fairbanks.*

an honorary certificate from the Musical Society in Brünn for his ongoing support of composers and musicians.[11]

Despite such recognitions and accomplishments, much of Mendel's abbacy was fraught with political turmoil, frustration, and disappointment. These difficulties began shortly after he assumed the abbacy in 1868. It was a politically tense time, not long after the transition of the Austrian Habsburg Empire to the Austro-Hungarian constitutional monarchy. The monarchy favored ethnic Germans and Hungarians over Czechs, a source of constant tension between Germans and Czechs in Brünn. Although Czechs constituted the majority in the province of Moravia, with Brünn as its capital, Germans held most public offices, and the German language was employed in official discourse.

Two opposing parties dominated politics of the monarchy: the Conservative Party, generally favored by ethnic Czechs and the Roman Catholic hierarchy, and the German Constitutional Party, also known as the Liberal Party, which controlled most elected offices and favored ethnic Germans and Hungarians. Within two years of his election as abbot, Mendel's preference for the Liberal Party had become well known, setting him at odds with the ethnic Czechs in the monastery and several of his church superiors. Mendel even signed a petition questioning the validity of the election of a Conservative Party member, publicly reaffirming his liberal political leanings.[12]

It would not be long, however, before Mendel found himself opposing members of the Liberal Party. A law passed in 1874 required a 10 percent annual tax on the value of monastery assets to pay members of the clergy through the state-sponsored religious fund. In 1875, Mendel received an order to send a valuation of the monastery property and assets, which he did. On this basis, the monastery was assessed an annual tax of 7,336 guilders for the period of 1875 through 1880, which for that time was excessively heavy.[13]

Mendel had both practical and philosophical reasons for opposing the law. First, when he assumed the abbacy, the monastery was already delinquent for ten years of back taxes, which were significantly less than the new tax on an annual basis. Although the monastery was wealthy in assets and real estate, it was cash poor due to unexpected yet essential expenses incurred during the latter years of Napp's abbacy. By Mendel's calculation, the new tax imposed a financial hardship that the monastery could not sustain. Second, even though the law had been passed and implemented according to constitutional procedure, Mendel firmly believed that it was unconstitutional. In response to the tax assessment, he sent what he called a "voluntary contribution" of 2,000 guilders, along with a protest of the tax, writing that this contribution was "the largest amount that the respectfully undersigned can withdraw from their establishment without doing it serious harm." He also noted in the letter that the monastery remained in debt for ten years of previous unpaid taxes, and this new tax would thus be an unbearable hardship.[14]

In response, the government returned the 2,000 guilders along with a renewed summons for the full tax. However, along with the summons was a notice that if the abbot believed the property and assets had been overassessed or that the ability of the members to carry out their duties would be compromised by the tax, he could make an appeal for a reduction. According to Iltis, the government had given Mendel a means to significantly reduce the tax, one that his fellow abbots utilized to "recognize the validity of the law and then to evade it."[15]

Mendel, however, remained firm in his resolve, refusing to appeal for a reduction and withholding any payment of the tax. He based his refusal, once again, on his claim that the tax was unconstitutional. The lord lieutenant of Moravia, who was responsible for administering the tax, forwarded Mendel's protest to his superior, the minister for public worship and education, with the statement that the opposition to the law, in Mendel's mind, was "an inexorable sense of duty" and was "based on a deplorable condition of mental tension."[16]

The minister replied to the lord lieutenant that "should the abbot remain recalcitrant, I leave it to Your Excellency to take the necessary legal measures to force compliance." The municipality of Brünn was charged with collection of the tax by serving a warrant and attempted to do so. However, the municipality reported to the lord lieutenant that "physical measures would have been needed to put the warrant in execution, the prelate having declared that the municipal officers would have themselves to take the keys out of his pocket if they wanted to open the cash-box, and would have to use force to remove any property for sale."[17]

Instead, a lien was placed against revenues generated by the monastery's holdings, such as the farms and dairies it owned. In announcing the lien to Mendel, the lord lieutenant sent a strongly worded reprimand, asserting that "the attempt to contest the legal validity of a constitutionally passed law and to represent the measures taken to carry this law into effect as unconstitutional conflicts grossly with the first and foremost duty of a loyal subject, which is to give absolute obedience to the existing laws, and least of all can such conduct be regarded with indifference in the case of the head of a monastery."[18]

Deduction of taxes from the monastery's revenues was compulsory from that time forward, and the state collected additional fees to compensate it for collection efforts. In the meantime, officials continued in their attempts to convince Mendel to change his mind. He first was offered honorary titles as an enticement, then later threatened with dismissal from some of the positions he held. Mendel consulted with attorneys, who also advised him to accede.

Instead, he continued to write formal protests against the law. He even went so far as to threaten the lord lieutenant that the monastery considered the state to be in debt for the funds collected under the lien and that, as abbot and curator of the monastery fund, he was recording 5 percent annual interest against the state. By 1877, all the taxes had been paid by compulsory collection, and an adviser to the lord lieutenant informed him that the lien could be withdrawn if Mendel agreed henceforth to comply with the law. However, the adviser reported that he had met with Mendel, who continued to affirm that "nothing will induce him to abandon his standpoint."[19]

Mendel's stubbornness morphed into additional threats. Late in 1878, he informed the lord lieutenant that the monastery would cancel its subscription to societies, suspend its contributions to charities, and publicize the evidence of how unjustly it had been treated. In his response on February 26, 1879, the lord lieutenant called Mendel's bluff, notifying him that he was welcome to publicize all records associated with the dispute, for they would show the public how serious his error of judgment had been.

By 1882, after additional protests, Mendel's church superiors felt forced to intervene, insisting that he comply with the law. Mendel countered that "it certainly remains difficult if not impossible to understand what right the bishops can arrogate to themselves in this matter . . . seeing that they have absolutely no jurisdiction."[20] Interestingly, among the records of this dispute is a handwritten document with tax information on one side and some notes Mendel made about *Hieracium* experiments on the back, suggesting that he was still thinking of his hybridization experiments during the time of the dispute or that he had used a leftover paper from the days of his experiments for recording notes.[21]

Mendel continued to lodge written protests, the final one on May 4, 1883. The following month, the lord lieutenant, by now weary of the ongoing resistance, appealed to the bishop of Brünn to again intervene on behalf of the church. Before any intervention could ensue, however, the procurator of the monastery informed the lord lieutenant that the abbot was too ill to deal with the matter and asked for leniency given Mendel's poor health. The procurator, apparently without Mendel's knowledge, sent a reevaluation of monastery property, deducting the value of the library collections, art, pensions, and other assets, resulting in a substantially lowered valuation. The minister for public worship and education agreed to the lowered valuation in November 1883 and made it retroactive for all previous years. Mendel by then was too ill to pay any attention to the matter, his death just a few weeks away. Following Mendel's death, this reassessment of the monastery's value resulted in a determination that the state had overtaxed the monastery during the period of the lien, and the amount of overpayment was returned to the monastery fund.

Gregor Johann Mendel passed away on January 6, 1884, at the age of sixty-one, after being bedridden with chronic heart problems and dropsy. He had been seriously ill since the previous summer. A postmortem examination, which Mendel had personally requested in the days preceding his death, showed that he suffered from kidney inflammation and heart failure.[22] Anticipating his passing, he penned his own death announcement, which was publicly released by the monastery:

> The Augustinian Monastery of St. Thomas at Altbrünn in Moravia respectfully and with profound regret informs the public of the death of the Right Reverend Abbot Gregor Joh. Mendel mitred prelate, companion of the Royal and Imperial Order of Franz Josef, emeritus chairman of the Moravian Mortgage Bank, member and one of the founders of the Austrian Meteorological Society, and various other learned and useful organizations, etc.[23]

Clemens d'Elpidio Janetschek, one of the youngest friars at the monastery who deeply loved his abbot, succinctly captured Mendel's passions and hardships in a few lines of Latin verse:

> *Clemens ac largus, affabilis unicuique,*
> *Fraternusque pater fratribus nobis fuit,*
> *Flores amavit et juris defensor vim toleravit,*
> *Qua tandem fessus, vulnere cordis obit.*
>
> Gracious, generous, and affable to everyone,
> Fraternal father to us brethren was he,
> Flowers he loved, and he fought to defend the law,
> Exhausted in the end, he died wounded in the heart.[24]

The funeral was held on January 9 in the Basilica of the Assumption of the Virgin Mary with every seat filled and a crowd of mourners outside. Although the Augustinians petitioned to have Mendel buried in the old cemetery, their request was denied, and his body was laid to rest in the sepulchers of the Augustinians in the Brünn central cemetery.[25] An ornate monument stands over the stone-capped vaults with the Latin inscription in red granite as the *Conventus Abbatialis* (Abbatial Assembly) (figure 9.3).

Figure 9.3. The Conventus Abbatialis monument standing at the site of the vaults where the remains of the abbots of the St. Thomas Monastery, including Mendel, are entombed. *Photograph by Daniel J. Fairbanks.*

All published accounts at the time of his death praised Mendel for a lifetime of contributions to science, education, the church, the region, and its people. In a Czech newspaper article, the author wrote, "The poor have lost in him a great benefactor, the friars a kind father, and many of us a genuine friend."[26] Although no one recognized the importance of his theory of inheritance, Mendel's scientific achievements were not neglected. He was praised as a successful horticulturalist, beekeeper, and meteorologist who had contributed much to the science of these disciplines. His political activism and obstinacy in the face of staunch opposition had won him friends and foes.

In the end, his death came with widespread respect, mingled with what must have been a sense of relief among his church superiors and government officials that the long taxation dispute had finally come to an end. No one at the time suspected that he would gain posthumous renown as one of history's greatest scientists, a triumph that would eclipse all accomplishments and notoriety he had at the time of his death.

Years later, Pater Janetschek recalled that Mendel's papers, presumably including the original data he tabulated for his experiments and his notes and his manuscripts for publication, were "committed to the flames."[27] Mendel's bound books, however, were retained in the monastery library. The notes penciled in those books, a few fragments of handwritten notes, and some letters escaped destruction, resurfacing years later in a variety of places. Some had been taken from their place of origin and distributed to various locations throughout the world during the late nineteenth and early twentieth centuries before their value was known. Others were destroyed during World War II and the communist occupation of Czechoslovakia. Disputes over their proper ownership are still fomenting. Lamentably, the vast majority of what Mendel recorded firsthand is forever lost.

The Rediscovery 10

The whole problem of heredity has undergone a complete revolution.

—WILLIAM BATESON AND EDITH REBECCA SAUNDERS

THE YEAR WAS 1900, "the turn of the century" as it is still known today. Sixteen years had passed since Mendel's death. The birth of the twentieth century stands as a milestone in human progress as stalwart traditions gave way to change. But change is rarely easy; resistance can be staunch and even violent. During the nineteenth century, with rare exceptions, women were neither expected nor allowed to choose scientific careers, often officially restricted from pursuing certain scholarly activities. As the twentieth century approached, opportunities for women in science were opening albeit slowly.

Such was the case at Charles Darwin's alma mater, Cambridge University. According to the rules, women could attend selected lectures, a practice started in the 1870s, but they could not receive degrees. In the final years of the 1800s, two Cambridge colleges for women, Girton and Newnham, were thriving. Even so, women still could not receive degrees; they were instead awarded less prestigious Tripos Certificates. Scottish universities by then offered degrees to women, but the administrators of Cambridge and Oxford held firm in their exclusivity favoring men.

In 1897, a group of 1,234 Tripos Certificate recipients of Girton and Newnham petitioned the Cambridge University Senate to retroactively grant them degrees. Massive protests erupted from large numbers of male students who vociferously opposed any change. They famously suspended two effigies of women overhanging Trinity Street. One was riding a bicycle, dressed in a puffy blouse, trousers, and knee-high striped stockings, hanging from a third-story window of the Macmillan and Bowes Bookshop. The other was a woman dangling upright from a line stretched between two buildings over the street. An enormous banner lampooning Shakespeare hung from the second story of Caius College, its words, in all caps, large enough to be read from hundreds of feet away: "GET YOU TO GIRTON BEATRICE, GET YOU TO NEWNHAM, HERE'S NO PLACE FOR YOU MAIDS, MUCH ADO ABOUT NOTHING." Hundreds of male students in boater and derby hats crowded into Trinity Street below the banner and effigies. When the vote to grant degrees to women failed—miserably—the men rioted in celebration, setting off fireworks and blasting confetti while vandalizing the university buildings.[1]

As the riots raged that May 1897, Edith Rebecca Saunders (figure 10.1) of Newnham College was tending young seedlings in the college's experimental garden from experiments she had initiated in 1895. In 1899, she had so distinguished herself as a scientist that she was named director of the Balfour Laboratory at Newnham. She had planned a set of elaborate hybridization experiments, similar to those Mendel had started almost four decades earlier. Unbeknownst to her, the experiments she was initiating that May would, within a few short years, solidify and expand Mendel's theory, elevating him to the highest echelon of scientific fame.

Figure 10.1. Edith Rebecca Saunders, whose experiments confirmed and amplified Mendel's theory, as she appeared in 1906. *Drawing by Daniel J. Fairbanks based on a historic photograph, sepia pastel on paper. Collection of the artist.*

Saunders was born in 1865, the year when Mendel presented his discoveries and theory. She entered Newnham College at Cambridge University the year of his death, in 1884, remaining there until her retirement as a teacher in 1925. Quiet and unassuming, her voice was silent when it came to political issues. Her passionate dedication to science mattered most. With single-minded attention to her research, she generated some of the most extensive and consequential discoveries reaffirming Mendel's theory. Respect for her contributions grew over the years. She was elected as a fellow of the Royal Horticultural Society and the Linnean Society, serving as vice president of the latter. She cofounded the Genetics Society in 1919 and later served as president of the Genetics Society in Great Britain from 1936 to 1938.[2] At the time of her death in 1945, the renowned mathematician and biologist J. B. S. Haldane eulogized her as one who "independently rediscovered some at least of Mendel's laws before his work was known."[3]

Sadly, little evidence is available to tell us about Saunders beyond her voluminous research. Aside from her scientific publications, only a few reminiscences, notes, letters, and photographs remain. As a botanist and geneticist, she dedicated her long and extraordinary life entirely to science and teaching.

Like many women scholars of the time, she relied out of necessity on a male mentor who was willing to sponsor her and coauthor her publications. Hers was William Bateson, a tall Cambridge don with a bushy moustache who dressed in tweed (figure 10.2). As a fellow of the Royal Society, he was coming into prominence, although he had no shortage of

Figure 10.2. William Bateson, Mendel's most vocal and visible advocate in the early twentieth century, a powerful supporter of women in science, and the scientist who coined the name genetics, as he appeared in 1906. *Drawing by Daniel J. Fairbanks based on a historic photograph, sepia pastel on paper. Collection of the artist.*

enemies, in part because of his active encouragement of women in science. Despite the resistance, he turned his attention to the women students of Newnham and Girton, recruiting them to participate in expanding the research Saunders had initiated.

Such mentorship was essential in the political climate of the time, but it also had its disadvantages, as stated by Wayne State University professor and historian Marsha Richmond:

> On a historiographic level, . . . the mentoring relationship could be a two-edged sword. Although mentors could facilitate the participation of women in academic science, such sponsorship could render them "invisible" when the women's contributions became associated with the mentor's name.[4]

Bateson, nonetheless, as Richmond described him, was "an avid supporter of the movement for the higher education of women."[5]

In early 1900, Saunders had assembled an enormous collection of experimental results and was analyzing and interpreting the data. Unbeknownst to her, other scientists in Germany, Austria, Holland, and the United States had independently conducted similar experiments, obtaining similar results. The spring of that year marked a dramatic series of independent and unplanned events that vaulted Mendel from obscurity to near-instantaneous fame. Biologists often refer to this sudden renaissance of Mendel's theory as "The Rediscovery," based on experiments independently carried out by scientists who observed the same patterns of inheritance Mendel had presented thirty-five years earlier. Who should be counted as one of Mendel's rediscovers remains a matter of conjecture, dependent on varying historical accounts and the criteria defining a "rediscoverer."

Regardless of who is considered as a rediscoverer, Bateson rapidly rose to the fore, leading a crusade to elevate Mendel's name to scientific renown. After reading Mendel's 1866 article for the first time in the late spring of 1900, he immediately began promoting Mendel with the passion of an evangelist, arguing vociferously in support of Mendel's theory in the face of formidable resistance. Five years later, he coined the term "genetics" to identify the new science, widely known by then as Mendelism.

Saunders was not one to openly voice her opinions—such actions from a woman scientist would be considered scandalous. In stark contrast, throngs of people overflowed the halls of Cambridge to hear Bateson publicly excoriate Mendel's detractors. Instead, as one of Bateson's star collaborators, Saunders and her colleagues provided him with the earliest experimental evidence he needed to promote the newly rediscovered Mendelian theory. Their collaboration rapidly grew into a multinational group of scientists, collectively self-named as the Mendelians.

How the rediscovery transpired is disputed, the various accounts littered with the sorts of anachronisms, contradictions, and uncertainties that are inevitable when people attempt to recollect distant events in the light of contemporary context. Mendel's rediscoverers are variously numbered as two, three, four, or more, depending on what is accepted as constituting "rediscovery" and what is considered as accurate history or discarded as apparent myth.

According to the traditional story, as first recounted by Bateson and Saunders,[6] three botanists, Hugo de Vries in Holland, Carl Correns in Germany, and Erich von Tschermak in Austria, were Mendel's original rediscovers (figure 10.3).

All three had an indirect connection to Mendel. De Vries received one of the rare reprints of Mendel's article in 1900, when a colleague sent it to him with the note, "I know that you are studying hybrids, so perhaps the enclosed reporting of the year 1865 by a certain Mendel, which I happen to possess, is still of some interest to you."[7]

Correns was Nägeli's student, recalling that "I had been made cognizant of Mendel's investigations through my teacher Nägeli. And I believe

Figure 10.3. Hugo de Vries (left), Carl Correns (center), and Erich von Tschermak (right) as they appeared in the first years of the twentieth century. They were botanists who independently carried out experiments that led to confirmation of Mendel's theory. All three published articles on Mendel in the spring and early summer of 1900. *Drawings by Daniel J. Fairbanks based on historic photographs, sepia pastel on paper. Collection of the artist.*

also to remember that he told me of Mendel, but certainly only of the *Hieracium* investigations, in which alone he was permanently interested."[8]

Tschermak's uncle was Eduard Fenzl, one of Mendel's two botany professors at the University of Vienna whose ideas Mendel discredited in his classic article. Tschermak, however, did not become aware of Mendel through his uncle. Instead, he first encountered Mendel's article, long after Fenzl's death, in a university library in Vienna while studying for his doctoral dissertation.

Like Mendel, Correns and Tschermak had chosen the garden pea as their experimental organism, both investigating contrasting character-pairs for seed shape and seed color, just as Mendel had, although their experiments were fewer and smaller than his. De Vries was working with other plant species, mostly ornamental flowers. Less known is the fact that Saunders also amassed a large set of experimental evidence on the offspring of plant hybrids in England at the same time de Vries, Correns, and Tschermak were conducting their experiments. The experimental results Saunders collected were some of the most extensive, and they would clarify and amplify Mendel's theory. Bateson was simultaneously experimenting with poultry, confirming Mendelian inheritance in animals.

The first to publish was de Vries, whose initial article was in French, titled "On the Law of Disjunction of Hybrids" ("*Sur la Loi de Disjonction des Hybrides*"), which appeared in April of that year.[9] The article did not attribute the newly discovered "law of disjunction" to Mendel, nor did de Vries mention Mendel in the article. He mailed a reprint to Correns, who received it on April 25. On reading it, Correns hurriedly penned a manuscript recounting his own experiments. By then, Correns had read Mendel's article and was anxious to attribute the new scientific law to him by name, starting with the article's title: "G. Mendel's Law on the Behavior of Progeny of Varietal Hybrids" ("*G. Mendel's Regel über das Verhalten der Nachkommenschaft der Rassenbastarde*").[10] On the first page, Correns emphatically established Mendel's priority with the following words (in English translation):

> When I discovered the regularity of the phenomena, and the explanation thereof—to which I shall return presently—the same thing happened to me which now seems to be happening to de Vries: I thought I had found *something new. But then I convinced myself that the Abbot Gregor Mendel in Brünn, had, during the sixties, not only obtained the same result through extensive experiments with peas, which lasted for many years, as did de Vries and I, but had also given exactly the same explanation, as far as was possible in 1866.* [emphasis in the original][11]

Another article by de Vries appeared soon after. It is in German with the title "The Segregation Law of Hybrids" ("*Das Spaltungsgesetz der Bastarde*"),[12] and in it, de Vries clearly attributes the theory of inheritance to Mendel.

In Vienna, Tschermak was a young doctoral student who had chosen hybridization of the garden pea for his doctoral dissertation, as yet unaware of Mendel. In January 1900, he presented his dissertation defense on his hybridization experiments, using seed shape and seed color as the contrasting characters, echoing Mendel's experiments. Like Correns, Tschermak received a reprint of the French article from de Vries in April. Having already read Mendel's article by this time, Tschermak determined to publish an article attributing the new scientific law to Mendel. By the time he received the printer's proofs, he had read the German articles by Correns and de Vries that mentioned Mendel, which he cited in the final version, published in June 1900.[13]

At this point, Bateson and Saunders enter the picture. Bateson was on a train in June 1900, traveling to a conference where he was scheduled to present the results of the latest research that Saunders and he had completed. Forty years later, well after Bateson's death, his spouse, Beatrice Bateson, recalled that he had an epiphany on the train when he read Mendel's paper for the first time:

> He had already prepared this paper, but in the train on his way to town to deliver it, he read Mendel's actual paper on peas for the first time. As a lecturer he was always cautious, suggesting rather than affirming his own convictions. So ready was he however for the simple Mendelian law that he at once incorporated it into his lecture.[14]

As is so often the case when historical accounts are amalgamated and retold, the stories of Mendel's rediscovery are littered with inconsistencies. Some historians have questioned, based on contradictions in the accounts of de Vries and Correns, which they recalled from memory decades later, whether they really had *rediscovered* Mendel's theory in the sense that they derived the same theory as Mendel entirely on their own *before* reading Mendel's article. Instead, one or both may have interpreted their results in a Mendelian fashion only *after* reading Mendel's article. There is strong evidence that de Vries thought in terms different than Mendel's theory before reading Mendel's article, that he may have read it as early as 1897 but did not fully understand it until early 1900.[15] There is definitive evidence that Correns read Mendel's article in 1896 but perhaps did not fully understand it at the time.[16] Tschermak, by his own account, had found and read Mendel's paper before he read the articles by de Vries and Correns and prepared his manuscript, although some have claimed that he did not understand it, even after reading the articles by de Vries and Correns, and therefore should be excluded from Mendel's so-called rediscoverers.[17]

Anachronisms in the account by Beatrice Bateson of her husband reading Mendel's article on the train suggest that he, in fact, read it after the supposedly epiphanous train ride and probably first learned of Mendel after reading the articles by de Vries and Correns.[18] Regardless of the exact moment when Bateson read Mendel's article, his enthusiastic acceptance and evangelistic promotion of Mendel were swift.

The historical evidence surrounding Saunders is so sparse that we cannot know when or how she learned of Mendel. By 1900, she had amassed an enormous body of experimental data confirming Mendel's theory. She and Bateson presented their findings in December 1901 and published them in early 1902.[19]

The research by Saunders soon attracted a cadre of students from Newnham and Girton who joined her and Bateson in the expanding research program, including Nora Darwin, who was Charles Darwin's granddaughter.[20] This effort firmly established the emerging science of Mendelism as one welcoming to women scientists, who initially constituted the majority of scientists engaged in it.

Reginald Punnett, who had recently graduated from Cambridge, would soon gain prominence with Bateson and Saunders as a leader in the new science. The famous Punnett square, often taught in biology classes, bears his name. Other men from Cambridge would likewise join the movement.[21]

On the European continent, Correns and Tschermak became staunch Mendelians, as did Lucien Cuénot in France, who discovered Mendelian patterns of inheritance in mice. In the United States, William Jasper Spillman had conducted similar hybridization experiments in wheat and reported observations of Mendelian proportions before learning of Mendel, though he described only consistency in numerical proportions without deriving a theoretical explanation. Spillman presented his results in a meeting in 1901, before learning of Mendel, and published them in 1902. Spillman's article was reprinted in a British journal in 1903, with commentary from Charles Hurst, who was one of Bateson's Mendelian associates.[22] Hurst's commentary on Spillman's article in 1903, after Mendel's article had been well publicized, connected Spillman's observations with Mendel's theoretical explanations.

Notably missing from the Mendelians in these early years was Hugo de Vries. In March 1900, he had written that Mendelism had "very general application." Just six months later, he reversed his view, his rejection evident in an often-quoted passage from a letter he wrote to Bateson on October 30, 1901:

> I prayed you last time, please don't stop at Mendel. I am now writing the second part of my book which treats of crossing, and it becomes more and more clear to me that Mendelism is an exception to the general rule of crossing. It is in no way *the* rule![23]

The dramatic rediscovery of Mendel unquestionably elevated his scientific status from obscurity to renown. However, like Darwin's theory, controversy surrounding Mendel's newly rediscovered theory was immediate and, at times, vitriolic, fueled more by a clash of personalities than a legitimate dispute over scientific interpretation. To understand the root of this controversy, we must go back in history to Mendel's time and to a debate surrounding Darwin's theory.

The debate, which by the time Mendel presented his classic theory had fomented for years, was about the nature of inherited variation. In some cases, variation was obviously discrete, such as distinct colored- or white-flowered pea plants and pigmented as opposed to albino fur in mammals. In other cases, variation was continuous, such as height in humans, with no distinct boundary between what might be considered tall or short. Mendel was fully aware of this fact, yet he intentionally excluded continuously varying characters that could not be easily and unambiguously distinguished from one another for the purpose of avoiding unintentional misclassification in his experiments. In his classic article, he wrote,

> Some of these characters, however, do not permit certain and sharp separation because the difference rests on a "more or less" that is difficult to determine. Such characters could not be used for the individual experiments, which had to be limited to characters that appear clearly and decidedly in the plants.[24]

Mendel, however, inferred (correctly as it turned out) that the theory of inheritance he derived from these discretely varying characters also applied to those that varied continuously:

> The perfect identity shown by all characters tested in the experiment fully permits and justifies the assumption that the same behavior applies to other

characters that appear less sharply in the plants and thus could not be included in the individual experiments.[25]

Not everyone, however, viewed discrete and continuous variation as manifestations of the same underlying process. Some argued that they were, in fact, quite different and that discrete Mendelian variation was the exception, not the rule. They typically viewed inheritance as being blended, like mixing differently colored paints to obtain a continuous range of intermediate colors, thereby explaining continuous variation. They argued that Mendel's theory promoted the idea of particulate inheritance, conferred by discrete hereditary elements. In turn-of-the-century Britain, disputes between the so-called continuarians and discontinuarians had reached a fevered pitch coinciding precisely with the Mendelian rediscovery. The time was ripe for a vicious war of words, with Mendel in the middle.

By 1900, self-appointed names for the two camps had become established. Those who adopted advanced mathematics as a means for studying continuous variation referred to themselves as "biometricians." Karl Pearson and Walter Frank Raphael Weldon (typically known by his initials, W. F. R.) led this camp and coined the term "biometrics" to describe their mathematical approach to biology. They even founded a new journal named *Biometrika* as a venue for publishing research on the topic. William Bateson and Hugo de Vries had been champions of discontinuity, referring to themselves and their followers as "discontinuarians." For example, Bateson, commenting on a meeting he had with de Vries, characterized him as "an enthusiastic Discontinuarian [who] holds the new mathematical [biometrical] school in contempt, so we hit it off to admiration."[26] After Mendel's theory had been rediscovered, the discontinuarians took on the name "Mendelians," pitting the biometricians against the Mendelians.

The biometricians considered themselves to be the followers of Francis Galton, Charles Darwin's half cousin, whom we encountered in chapter 7. Ironically, Galton had invented the term "particulate inheritance," the alternative to blending inheritance promoted by the biometricians. Galton correctly proposed that particulate inheritance could explain the appearance of continuous variation in populations and among the offspring of differing parents, as is evident in the following statement about the inheritance of skin pigmentation in humans: "It need be none the less 'particulate' in its origin, but the result may be regarded as a fine mosaic too minute for its elements to be distinguished in a general view."[27] Neither the biometricians nor the Mendelians denied the simultaneous existence of both types of variation, continuous and discontinuous, in nature. The two camps were instead divided on the relative importance of each.

Weldon had at one time been Bateson's close friend and his mentor at Cambridge. But disagreements in the 1890s turned them into the most bitter of enemies. After Bateson and Saunders published their seminal article on Mendel in January 1902, Weldon published an article titled "Mendel's Laws of Alternative Inheritance in Peas" in the newly minted first volume of the journal *Biometrika*.[28]

Here Weldon criticized Mendel in two respects. First, he noted that Mendel's experimental results were remarkably close to those he predicted. Although Weldon did not overtly claim that Mendel altered his experimental results to match his theory, he conspicuously dropped several hints, such as "some writers have been led to overlook the wonderfully consistent way in which Mendel's results agree with his theory. . . . Here again, therefore, Mendel's statement is admirably in accord with his experiment," and "the odds against a result as good as this or better are 20 to 1."[29] Second, Weldon argued that Mendel's theory was limited and that even the character-pairs Mendel studied in peas did not, according to Weldon, consistently follow the laws Mendel proposed. Weldon relied mostly on observations that seed color and seed wrinkling in peas appeared to vary in a continuous fashion in some pea varieties and in their offspring, and he provided examples. In the end, Weldon did not hide his disdain for Mendel's approach:

> The fundamental mistake which vitiates all work based upon Mendel's method is the neglect of ancestry, and the attempt to regard the whole effect upon offspring, produced by a particular parent, as due to the existence in the parent of particular structural characters; while the contradictory results obtained by those who have observed the offspring of parents apparently identical in certain characters show clearly enough that not only the parents themselves, but their race, that is their ancestry, must be taken into account before the result of pairing them can be predicted.[30]

Bateson was incensed! In his mind, Weldon's article was an unpardonable attack, requiring a vigorous rebuttal. He quickly published a booklet of 102 pages in response, titled *Mendel's Principles of Heredity: A Defence.*[31] About half of the booklet consists of the first English translations of Mendel's two most significant articles, the classic 1866 article on peas and the lesser-known 1870 hawkweed article. In the remaining half, Bateson vigorously defends Mendel. Bateson's quasi-religious admiration for Mendel and his contempt for Weldon are evident in the preface:

> But every gospel must be preached to all alike. It will be heard by the Scribes, by the Pharisees, by Demetrius the Silversmith, and the rest. Not lightly do men let their occupation go; small then, would be our wonder, did we find the established prophet unconvinced. Yet, is it from misgiving that Mendel had the truth, or merely from indifference, that no naturalist of repute, save Professor Weldon, has risen against him?
>
> In the world of knowledge, we are accustomed to look for some strenuous effort to understand a new truth even in those who are indisposed to believe. It was therefore with a regret approaching to indignation that I read Professor Weldon's criticism. . . .
>
> If I have knocked a trifle loud, it is because there is need.[32]

In the chapters following the English translations of Mendel's articles, Bateson presented an excruciatingly detailed rebuttal to Weldon's article, offering counterarguments point by point, buttressed with exhaustive references to earlier published works. Bateson then completed his booklet with a passage drenched in sarcasm:

> Professor Weldon declares he has "no wish to belittle the importance of Mendel's achievement"; he desires "simply to call attention to a series of facts which seem to him to suggest fruitful lines of inquiry." In this purpose I venture to assist him, for I am disposed to think that unaided he is—to borrow Horace Walpole's phrase—about as likely to light a fire with a wet dish-clout as to kindle interest in Mendel's discoveries by his tempered appreciation.[33]

George Udny Yule, a statistician with an interest in Mendelism, perceived the conflict as fueled more by personal animosity than by any scientific incompatibility. As the war of words between the biometricians and Mendelians heated up, Yule attempted to mediate. In a lengthy review, Yule acknowledged that Bateson could "congratulate himself on a *succès de scandale*" but warned that "his speculations would have had more value had he kept his emotions under better control; the style and method of the religious revivalist are ill-suited to scientific controversy."[34] After explaining the compatibility between the two approaches, Yule claimed that "Mendel's Laws [promoted by Bateson] and the Law of Ancestral Heredity [promoted by Weldon] are not necessarily contradictory statements, one or other of which must be mythical in character, but are perfectly consistent the one with the other and may quite well form parts of one homogeneous theory of heredity."[35] Although Yule's conclusion would soon gain support from experimental evidence, his attempt at mediation utterly failed to calm the wrath of either Bateson or Weldon.

In 1902, Bateson traveled to New York, where the Royal Horticultural Society held its second conference on hybridization, and the theme of the conference was Mendel. The collective effort of the Mendelians amassed some of the finest science of those years. Bateson's followers documented numerous instances of Mendelian ratios in diverse species of animals and plants, adding abundant and irrefutable evidence to bolster Mendelism. In England, Saunders and Punnett continued working directly with Bateson researching Mendelian inheritance in numerous plant and animal species, coauthoring several of the most important articles with him.

One of the most significant claims made by Bateson was the assertion that Mendelism applied to humans. After studying the observations of Archibald Garrod, a British physician, on the inheritance of a human hereditary disorder known as alkaptonuria (darkening of the urine on exposure to the air), Bateson and Garrod agreed that it must be a recessive Mendelian character. Garrod published a large collection of data from human pedigrees supporting this determination.[36] Over the next several years, other genetic disorders in humans displaying either recessive or dominant Mendelian inheritance were documented in human pedigrees. Bateson sought out the examples and cataloged them, firmly establishing Mendelism as the mechanism of human inheritance.

In the meantime, deeply wounded by Bateson's public attacks, Weldon became obsessed with attacking Mendelism. He took on a student named Arthur Dukinfield Darbishire, who began experiments with peas and mice at Weldon's direction. Under the influence of his mentors, Weldon and Pearson, Darbishire initially interpreted his results as non-Mendelian. However, unlike Weldon, who by then had severed all private communication with Bateson, Darbishire freely wrote personal letters to Bateson, in them sharing his experimental results prior to publication. In one of these letters, Darbishire wrote, "I am absolutely unbiased about Mendel and am very keen to come to an unprejudiced conclusion on it."[37] Darbishire had naively offered himself as a mole, sending Bateson the information he and Weldon had obtained prior to publication. He also became a pawn whom Bateson would not hesitate to play in his spiteful match against Weldon.

Bateson was highly interested in Darbishire's experiments and corresponded with him regularly. In each case, Bateson reinterpreted Darbishire's experiments in a Mendelian fashion. Weldon, in turn, accused Bateson of revising Mendel's theory to suit his needs rather than admitting its flaws. In an article titled "Mr. Bateson's Revisions of Mendel's Theory of Heredity," Weldon wrote, in reference to Bateson's interpretation of Darbishire's experiments, "Mr. Bateson has stated his belief that the inheritance of eye-colour in these mice is 'strictly Mendelian,' and he has published a Mendelian formula which implies his belief that the inheritance of coat-colour is Mendelian also."[38] After arguing that Bateson had failed in this attempt, Weldon issued a public challenge to Bateson: "I earnestly appeal to Mr. Bateson, now that the facts are fully before him, to produce some final formula, expressed in terms of categories so definite that they really describe the mice included, which seems to him capable of bringing the behaviour of these hybrid mice and their offspring into harmony with the 'principles' he so strongly supports."[39]

Bateson relished Weldon's challenge and was ready for it. His arguments were so convincing that he began to win Darbishire to his side. Within a year, after consistently observing Mendelian ratios in his experiments, Darbishire was fully convinced of Mendelism while he was still Weldon's student, putting him in an untenably awkward predicament. To add fuel to the fire, Bateson was ready to publish the errors he had found in Darbishire's interpretations, and he had informed Darbishire of this. In desperation, Darbishire wrote to Bateson, "I hope you will do your best to get me out of the position I am in as soon as possible and I pray you not to mention this letter to anyone."[40] By the following year, 1904, Darbishire was fully on board with the Mendelians to the chagrin of Weldon and Pearson, both of whom expressed their disappointment with him in print.

The dispute between the Mendelians and biometricians came to a head in a meeting of the Zoology Section of the British Association held on August 18 and 19, 1904. Recalling the events of the most contentious session of that meeting, Punnett reminisced,

> We adjourned for lunch and on resuming found the room packed as tight as it could hold. Even the window sills were requisitioned. For word had gone round that there was going to be a fight. . . . Toward the end Pearson got up and the gist of his remarks was to propose a truce to controversy for three years, after which the protagonists might meet again for further discussions. On Pearson returning to his seat, the Rev. T. R. Stebbing, a mild and benevolent looking little figure for a great carcinologist, rose to conclude the discussion. In a preamble he deplored the feelings that had been aroused, and assured us that as a man of peace such controversy was little to his taste. We all began fidgeting at what promised to become a tame conclusion to so spirited a meeting, especially when he came to deal with Pearson's suggestion of a truce. But we need not have been anxious, for the Rev. Mr. Stebbing had in him the making of a first-rate impresario. "You have all heard," said he, "what Professor Pearson has suggested" (pause), and then with a sudden rise of voice, "But what I say is let them fight it out."[41]

Predictably, the flames of animosity between Bateson and Weldon raged even hotter. Experimental evidence, however, was accumulating in favor of the Mendelians—hundreds of experiments in plant and animal species had displayed Mendelian patterns, and the number of scientists conducting hybridization experiments was increasing. Many of the apparent exceptions were fully explainable as extensions of Mendelian principles, as Bateson and his collaborators deftly pointed out in one publication after another and as Mendel himself had anticipated, observed, and explained in his original article.

Yule again weighed in as a mediator, this time on one of the major issues in the dispute: the relative importance of continuous and discontinuous variation. According to Yule, the inheritance of continuous variation could be explained with Mendelian principles if one presumed that many hereditary elements, each inherited in a Mendelian fashion, influenced the same trait, so that any one pair of elements had only a minor influence, but collectively their influence would result in the appearance of continuous variation.

It did not take long for Yule's hypothesis to be confirmed experimentally. A Danish Mendelian, Wilhelm Johannsen, published a groundbreaking book in German titled *Elements of the Exact Theory of Heredity* (*Elemente der exakten Erblichkeitslehre*), wherein he coined the word "gene" to designate those hereditary elements as discrete underlying units of inheritance.[42] Close to the same time, Herman Nillson-Ehle published his doctoral dissertation at the University of Lund in Sweden on kernel color in wheat, providing solid evidence that three pairs of genes inherited in a Mendelian fashion influenced the amount of red color in wheat kernels in an apparently continuous fashion.[43] Edward East, a Harvard geneticist, published a series of experiments on kernel color in maize[44] and later with flower length in tobacco.[45] East's and Nillson-Ehle's experiments conclusively confirmed what Yule had proposed: both continuous and discontinuous variation had a firm basis in Mendel's theory, just as Mendel himself had inferred.

Weldon would not live to see that day, however. In April 1906, Weldon collapsed with illness and suddenly died. His family and friends harbored animosity toward Bateson for the long dispute that had taken so much of Weldon's time and energy and their suspicion that the stress from it contributed to his premature death. Pearson did not pursue the fight after Weldon's passing, bringing the conflict between the biometricians and the Mendelians to a somber end.

During this tragic controversy, Bateson had written a letter to Alan Sedgwick in 1905 in which he first used the word "genetics." The term first appeared in public, repeatedly and extensively, in the *Report of the Third International Conference 1906 on Genetics.*[46] In a ceremony at the conference, seven medals were awarded, one of them to Saunders, one of the first women in the history of science to be so honored.[47] The citation stated, "Miss Saunders has conducted the most intricate and difficult researches on the basis of Mendel's laws—researches demanding the utmost exercise of patience, coupled with the keenest observation."[48] From that time forward, the name "genetics" would remain firmly established to define the newly rediscovered science founded by Mendel.

Epilogue
Tragedy and Heroism

At twilight on March 8, 2015, I stood in a lovely stone-paved plaza in Brno, facing the building that in Mendel's day was the *Realschule*. One hundred and fifty years earlier, in the evening of that same date in that same building, Mendel presented his theory to his scientific colleagues. As I stood there, the sound of bells ringing in the building's campanile pierced the chilled air. After traversing the plaza several times, from one place to another, I found the vantage site I had been seeking, one where I could view the entire building unobstructed by nearby modern structures. I stood there, the cold air stiffening my hand movements, and made a small freehand sketch in ink on Czech laid paper (figure E1). Later, in my studio, I made a more refined drawing based on that sketch, shown in figure P1 of this book's prologue.

Dozens of people walked by, oblivious to what I was doing or the meaning of that moment in history. I had just finished attending a two-day symposium commemorating the sesquicentennial of Mendel's March 8, 1865, presentation in the nearby Moravian Museum.[1] The symposium consisted of two days of presentations by scholars from throughout the world; I was honored to be one of the presenters. The previous evening, we had toured Brno on foot, visiting sites important to Mendel, guided by Anna Matalová, one of the world's most knowledgeable Mendel scholars and emeritus head of the Mendelianum. As I stood there sketching the building, my mind mused on the wealth of information my colleagues and I had shared with one another over the past two days, details of rich Mendelian history spanning a century and a half. This plein air sketch was my personal epilogue to an extraordinary event.

Figure E1. Freehand sketch, made in the evening of March 8, 2015, of the building where Mendel presented his theory 150 years earlier on that same date. *Drawing by Daniel J. Fairbanks, sepia ink on Czech laid paper. Collection of the Mendelianum, Moravian Museum, Brno.*

One of the presenters at the symposium that day informed us of another symposium held fifty years earlier, in 1965, commemorating the centennial of Mendel's 1865 presentations. At the time, Brno was in communist Czechoslovakia, part of the Warsaw Pact nations. Friars no longer roamed the halls where Mendel had once been abbot, the communist regime having seized the building in 1950, arresting its Augustinian occupants.[2] The 1965 symposium itself was in peril. As Vítězslav Orel recalled,

> The very fact that Mendel was not only a geneticist but also a Catholic prelate raised a question about the celebration in Czechoslovakia, and attracted the interest of Antonín Novotný, then first secretary of the Communist party and President of the Republic. It was not until later that I learned that Khrushchev, when he visited Czechoslovakia in 1964, had rebuked Novotný for allowing the organization of the Mendel symposium in Brno.[3]

Fortunately, that 1965 symposium proceeded without incident, its crowning event the inauguration of the Mendelianum, a division of the Moravian Museum dedicated to Mendel, with space granted to it on the ground floor of one of the monastery buildings where Mendel had lived most of his life. Jaroslav Kříženecký, head of the Gregor Mendel Department of Genetics of the Moravian Museum, had the vision and determination to establish the Mendelianum. His death in December 1964, a few months before the commemorative symposium, deprived him of the honor of witnessing the inauguration of the museum, which he had spent much of his life striving to establish. Shortly before his passing, he reminisced on the wonder that such an honor to Mendel could even happen: "It was a paradox of history that this was achieved under conditions that permitted Mendel to be humbled and discredited to an extent that only scholars of history can appreciate."[4]

These "conditions that permitted Mendel to be humbled and discredited" comprised one of the twentieth century's most tragic periods, characterized by University of California, Berkeley, geneticist Michael Lerner as "the most bizarre chapter in the history of modern science."[5] It began with Joseph Stalin's purge of Soviet intellectuals, coupled with the banishment of Mendelian genetics under penalty of imprisonment and even death. Kříženecký was one of several geneticists who suffered imprisonment yet survived to recount the tragedy—and the heroism—of that time.

The story begins with a Soviet botanist named Nicolai Vavilov. Enamored by the eloquence of Mendel's theory and the abundant evidence supporting it, Vavilov traveled to England to study genetics under William Bateson in 1913. On his return to his native country the following year, he founded the study of genetics in Soviet Russia as a firm devotee of Mendel's theory and one of Bateson's growing number of protégés. As a plant breeder, he applied Mendelian principles to develop new, highly productive varieties of essential food crops, advancing Soviet agriculture. He is best remembered, however, for his epic journeys to some of the most remote regions of the world to collect and preserve the rich genetic diversity of native heirloom varieties of food plants—grains, vegetables, potatoes, beans, peas, peanuts, fruit trees and vines, squash, melons, and many more. The genetic diversity of those varieties was in peril of being lost through expanding modernization. His effort eventually grew into a worldwide endeavor to preserve the diversity of domestic plants and animals created through human ingenuity and cultivation over millennia. He and his followers amassed thousands of native heirloom varieties, cataloging and preserving them at an institute in Leningrad (now St. Petersburg), in what is called a gene bank, like priceless treasures in a museum. By the early 1930s, his star was rising as a world-renowned Mendelian and as a Soviet patriot.

Beginning in the late 1920s, Trofim Lysenko, a relatively unknown agricultural worker in Soviet Azerbaijan, implausibly claimed that he had discovered methods to phenomenally increase agricultural production in ways that directly contradicted Mendel's theory. According to him, "You need but change the type of metabolism in a living body to bring about a change in heredity."[6] Lysenko had redefined an old and discredited assertion known as "the inheritance of acquired characteristics," also known as Lamarckism because the French naturalist Jean Baptiste Lamarck had promoted the notion more than a century earlier. As Lysenko claimed, changes in the environment could mold hereditary change, a subject Mendel had addressed and dismissed in his classic article. The newly resurrected and revised Soviet version of Lamarckism soon became known as Lysenkoism.

Attracting the attention and approval of Stalin, Lysenko and his growing number of followers forced political ideology into science through their proclamations that Soviet geneticists, led by Vavilov, had fallen prey to "bourgeois opinions in science." Lysenko's chief spokesperson, Isaak Prezent, writing about Mendelian geneticists, proclaimed that "the philosophical foundations of the theory [of Mendel] they defend had already found a place in the history of pseudoscience."[7]

Vavilov refused to remain silent. He directly challenged Lysenko in heated public debates and discredited him in his writings. Lysenko and Prezent, in response, named Vavilov as an "enemy of the people" in 1937.

As political winds in the Soviet Union turned against those who accepted Mendel's theory, Vavilov boldly uttered in 1939 what are now his famous foreboding words, quoted in the this book's prologue: "We shall go to the pyre, we shall burn, but we shall not retreat from our convictions."[8]

Vavilov was true to his word in never retreating from his convictions. In 1940, he was arrested and subjected to a brief trial. The three-membered court convicted him on numerous counts, among them "belonging to a rightist conspiracy; spying for England; . . . sabotage in agriculture; [and] links with white émigrés."[9] The court sentenced him to death, though the sentence was not immediately carried out.

The Nazi army, by then, was advancing into the Russian homeland. Vavilov and his fellow prisoners were transported away from the war front to Saratov Prison. In the meantime, his colleagues at the institute in Leningrad quietly continued his legacy, caring for the precious stores of seeds he and they had collected. As the frigid winter of 1941–1942 descended on Leningrad, the Nazi army surrounded the city, cutting it off from food supplies and starving tens of thousands of its residents to death. The seed collections in the institute were much too small to feed even a few people, but they could have spared the curators there—if they chose to eat them. Instead, recognizing the value of the collections for the future of humanity, they took an oath to preserve them, even at the peril of their own lives. Nine of them made the ultimate sacrifice, dying of starvation while keeping their oath, surrounded by carefully cataloged packets of seeds. One of the survivors, Vadim Lekhnovich, recalled the severity of the ordeal and their steadfast commitment:

> It was hard to walk. . . . It was unbearably hard to get up every morning, to move your hands and feet. . . . But it was not in the least difficult to refrain from eating up the collection. For it was *impossible* [to think of] eating it up. For what was involved was the cause of your life, the cause of your comrades' lives.[10]

Vavilov died in prison of malnutrition and disease in 1943. He was one of many scholars who suffered such a tragic fate under Stalin's purge.

After World War II, Lysenko's power grew, as did his vitriol against those whom he and his followers derogatorily called Mendelists. In 1948, the Soviet Communist Party seized the government in Czechoslovakia and suppressed Mendel's legacy in his homeland. Křížencký was dismissed from his university position in 1949 for being a "reactionary Mendelist."[11] The communist secret police closed the St. Thomas Monastery in 1950 and arrested the Augustinian friars who resided there.[12] Mendel's greenhouse was demolished, and the monastery fell into disrepair. Křiženecký was later arrested and imprisoned for a year and a half in 1958 for publishing an article supportive of Mendel and critical of Lysenko.[13]

According to Vítězslav Orel, who would later lead the Mendelianum and write a definitive biography of Mendel, "In the same year (1958) I was dismissed from the research institute and was not allowed to have any kind of appointment in Brno. It seemed clear to me that all those dreams for a Mendel Institute of Genetics in Brno would not be realized, at least not in the near future."[14]

As the 1960s approached, biologists throughout the world increasingly condemned Lysenko's totalitarian rule over biology as Soviet dissidents revealed the many atrocities that had taken place. A 1962 Soviet commission determined to censure Lysenko, but at the last minute, the commission was disbanded at the direction of Nikita Khrushchev, first secretary of the Communist Party of the Soviet Union. Khrushchev then publicly reaffirmed his support of Lysenko. In 1963, at great risk, Zhores Medvedev and Valentin Kirpichnikov succeeded in publishing an article in the Soviet magazine *Neva*, condemning Lysenkoism, its publication made possible through the bravery of the magazine editors:

> How, then, did it happen that in our country the development of genetics was so long delayed? Why did we for so long yield to the capitalist countries this large and productive area of the scientific front under the excuse that genetics is a bourgeoise science? . . . The answers to the questions posed seem to us to be very simple. It could have happened only in the environment of distortions observed in the era of personality cult. The attempt to isolate Soviet biology from world science is a harmful remnant of the personality cult, an alienation from reality, a fear of openly and honestly admitting and correcting previous errors.[15]

As Lysenko's reputation rapidly crumbled, Khrushchev continued his unabated support. But the tide was turning. Soviet scientists from fields beyond biology mustered the courage to join the biologists in their collective resistance to Lysenko. One of the most vocal was the famed nuclear physicist Andrei Sakharov.

The tragedies and embarrassment of Lysenko's long and tortuous reign over Soviet biology were among the many factors contributing to Khrushchev's ultimate downfall in October 1964. Lysenkoism precipitously collapsed soon thereafter, with Lysenko's abrupt dismissal in February 1965.

At this same time in Brno, Křiženecký was in the final throes of a painful battle with cancer. According to Orel, "Křiženecký died in December

1964 and did not live to see the triumph of the science to which he had devoted his life."[16] Reminiscing on the events of that time, Orel recalled how Křiženecký convinced him to alter the course of his career to reestablish Mendel's legacy in his homeland:

> Toward the end of 1963 . . . I was hoping to be allowed to study genetics and to engage in experimental research. Křiženecký convinced me in 1964 to accept this completely different kind of appointment, however, by explaining that our first task was to rehabilitate Mendel as a man and a scientist, and at the same time to free genetics from its subordinate position to ideology.[17]

In August 1965, biologists and historians from throughout the world assembled in Brno for the symposium, mentioned at the beginning of this epilogue, commemorating the centennial of Mendel's presentations. Honoring Křiženecký's legacy and request, Vítězslav Orel took the helm of the newly founded Mendelianum. With funding from UNESCO, a prestigious scholarly journal, the *Folia Mendeliana*, was established as a venue for scholars to publish their research on Mendel's history and his legacy. I am honored to have published some of my own research within its pages.

As mentioned in this book's preface, in 1993, I made the pilgrimage every geneticist yearns to make—to Mendel's monastery and the place where he lived and cultivated his experimental plants. Earlier that year, Czechoslovakia had been dissolved and the Czech Republic formed. A mere three and a half years had passed since the 1989 Velvet Revolution, when the communist regime in Czechoslovakia peacefully fell, replaced by a parliamentary republic. Shortly after that revolution, the new government returned the St. Thomas Monastery to the Augustinian Order of the Roman Catholic Church. I distinctly recall viewing the sets of cloth slippers in worn wooden compartments belonging to the friars who had recently reoccupied the monastery where Mendel and his brethren had once roamed the halls.

Anna Matalová, head of the Mendelianum, took several hours out of her busy schedule to guide us on a tour of the museum, the grounds, and Mendel's bee house. She obtained permission from the Augustinians for us to enter the library and view its ornate furnishings and ceiling paintings. Importantly, she retrieved Mendel's personal copies of books by Darwin and Gärtner so that I could photograph his annotations in those books, which are highlighted in several images in this book.

I shall forever remember an exhibit on the wall of the museum, not far from where the "Great Prelate Portrait" (see figure 9.1) was hanging, listing the names of scientists who were imprisoned and died under Lysenkoism. Dr. Matalová recounted some of the sacrifices of those scientists, emotion obvious in her voice. She then told me of her experiences at greeting the former students of these fallen heroes. They were finally free to travel to the place where Mendel conducted and presented his experiments under their newfound liberty with the fall of communism. In solemn tribute, they each gathered a bit of soil from Mendel's garden to take home and scatter on the graves of their mentors.

The two centuries since Mendel's birth have marked the life, legacy, controversies, tragedies, and triumphs of one of history's most influential scientists. His theory, in its clarity and mathematical elegance, continues to serve as an enduring foundation for biology and as a timeless paradigm of experimental science.

Appendix
Experiments on Plant Hybrids by Gregor Mendel

Submitted at the meetings of February 8 and March 8, 1865

English translation by Scott Abbott and Daniel J. Fairbanks
Originally published in *Genetics* 204 (2016): 407–22
https://www.genetics.org/content/204/2/407

Introductory Remarks

Artificial fertilisations of ornamental plants to produce new colour variants led to the experiments to be discussed here. The striking regularity with which the same hybrid forms reappeared whenever fertilisation took place between the same species was the stimulus for further experiments, whose objective was to follow the development of hybrids in their progeny.

Careful observers like *Kölreuter, Gärtner, Herbert, Lecocq, Wichura and others* have tirelessly sacrificed parts of their lives to this objective. Gärtner especially, in his work "The Production of Hybrids in the Plant Kingdom," documented very worthwhile observations, and most recently, Wichura published fundamental researches on willow hybrids. That a generally standard law for the formation and development of hybrids has not yet been successfully given is no wonder to anyone who knows the extent of the subject and who realises the difficulties with which experiments of this kind must struggle. A final determination will result only when *detailed experiments* on the most diverse plant families are available. Anyone who surveys the work in this area will be convinced that among the numerous experiments, none has been carried out in the extent and manner that would make it possible to determine the number of the various forms in which the progeny of hybrids appear, so that one could, with confidence, arrange these forms into the individual generations and determine their relative numerical relationships. Some courage is certainly required to undertake such an extensive work; nevertheless, it seems to be the only proper means to finally reach resolution of a question regarding the evolutionary history of organic forms, the importance of which must not be underestimated.

The present treatise discusses an attempt at such a detailed experiment. It was, as the task required, limited to a relatively small group of plants and was essentially completed only after the course of eight years. Whether the plan by which the individual experiments were arranged and carried out corresponds to the given objective, that may be determined through a benevolent judgment.

Selection of the Experimental Plants

The worth and validity of any experiment is determined by the suitability of the materials as well as by their effective application. In this case as well it cannot be unimportant which plant species are chosen for the experiment, or the manner in which it is conducted.

The selection of the group of plants for experiments of this kind must be done with the greatest care if one does not wish to put the results in question from the beginning.

The experimental plants must necessarily

1. Possess constantly differing characters.
2. At the time of flowering, their hybrids must be protected from the action of all pollen from other individuals, or be easily protected.
3. The hybrids and their progeny in the succeeding generations must not suffer any noticeable disturbance in fertility.

Adulteration through pollen from another individual, if such were to occur unrecognised in the course of the experiment, would lead to completely false conclusions. Impaired fertility or complete sterility of individual forms, like those that appear in the progeny of many hybrids, would greatly impede the experiments or thwart them completely. In order to recognise the relationships of the hybrid forms to one another and to their original parents, it appears to be necessary that *every* member that develops in the series in every single generation be subjected to observation.

From the beginning, special attention was given to the *Leguminosae* because of their curious floral structure. Experiments made with several members of this family led to the conclusion that the genus *Pisum* sufficiently meets the necessary requirements. Several completely independent forms of this genus possess uniform characters that are easily and certainly distinguishable, and they give rise to perfectly fertile hybrid progeny when reciprocally crossed. Disturbance by pollen from other individuals does not easily occur, as the organs of fructification are tightly enclosed by the keel and the anthers burst early in the bud so that the stigma is covered by pollen before the flower opens. This circumstance is of special importance. Other advantages that deserve mentioning are the ease of cultivating these plants in open ground and in pots, as well as their relatively short vegetative period. Artificial fertilisation is, no doubt, somewhat laborious, but it is almost always successful. For this purpose, the not yet perfectly developed flower bud is opened, the keel separated and each stamen slowly removed with forceps, whereupon the stigma can immediately be dusted with pollen from another individual.

A total of thirty-four more or less different pea varieties were obtained from several seed suppliers and subjected to a two-year trial. In one variety a few greatly distinct forms were noticed among a larger number of identical plants. The next year there was no variation among them, however, and they matched another variety obtained from the same seed supplier in every way; without doubt the seeds had been accidentally mixed. All the other varieties produced absolutely identical and constant progeny; at least in the two trial years no essential variation was noticed. From these, 22 were selected for cross-fertilisation and were cultivated annually throughout the duration of the experiments. Without exception they held true to type.

The systematic classification is difficult and uncertain. If one were to apply the strictest definition of the term species, according to which only those individuals that display precisely the same characters under precisely the same conditions belong to the same species, then no two could be counted as a single species. According to the opinion of experts in the field, however, the majority belong to the species Pisum sativum, while the others were considered and described as sub-species of P. sativum, sometimes as independent species, such as P. quadratum, P. saccharatum, P. umbellatum. In any case, these systematic ranks are completely unimportant for the experiments described here. It is as impossible to draw a sharp line of distinction between species and varieties as it is to establish a fundamental distinction between the hybrids of species and varieties.

Arrangement and Order of the Experiments

If two plants that are constantly different in one or more characters are united through fertilisation, the characters in common are transmitted unchanged to the hybrids and their progeny, as numerous experiments have shown; each pair of differing characters, however, unite in the hybrid to form a new character that generally is subject to variation in the progeny. To observe these variations for each pair of differing characters and to ascertain a law according to which they occur in succeeding generations was the objective of the experiment. This experiment, therefore, breaks up into just as many individual experiments as there are constantly differing characters in the experimental plants.

The different pea forms selected for fertilisation show differences in the length and colour of the stem, in the size and form of the leaves, in the placement, colour and size of the flowers, in the length of the flower peduncles, in the colour, form, and size of the pods, in the form and size of the seeds, in the colour of the seed coat and of the albumen. Some of these characters, however, do not permit certain and sharp separation because the difference rests on a "more or less" that is difficult to determine. Such characters could not be used for the individual experiments, which had to be limited to characters that appear clearly and decidedly in the plants. A successful result would finally show whether they all are observed as portraying identical behaviour in hybrid union and whether, as a result, a judgment is possible about those characters that typically are inferior in their importance.

The characters included in the experiments relate to:

1. The *difference in the form of the ripe seeds.* These are either spherical or somewhat rounded, the depressions, if any occur on the surface, are only shallow; or they are irregularly angular and deeply wrinkled (P. quadratum).
2. The *difference in the colour of the seed-albumen* (endosperm). The albumen of the ripe seeds is either pale yellow, bright yellow or orange coloured; or it possesses a more or less intensive green colour. This difference in colour is obvious to see in the seeds, since their coats are translucent.
3. The *difference in the colour of the seed coat.* This is either coloured white, a character consistently associated with white flower colour; or it is grey, grey-brown, leather-brown with or without violet spots, in which case the colour of the standard petal appears violet, that of the wings purple, and the stem at the base of the leaf axils

is tinged reddish. The grey seed coats turn blackish-brown in boiling water.

4. The *difference in the form of the ripe pod.* This is either simply inflated, never pinched in places, or it is deeply constricted between the seeds and more or less wrinkled (P. saccharatum).
5. The *difference in the colour of the unripe pod.* It is either light- to dark-green or coloured a bright yellow, a colour shared by stems, leaf veins, and sepals.[1]
6. The *difference in the placement of the flowers.* They are either axial, i.e. distributed along the stem, or they are terminal, accumulated at the end of the stem in a short false umbel; in which case the upper part of the stem is more or less widened in cross-section (P. umbellatum).
7. The *difference in the length of the stem.* The length of the stem is very different in individual forms, however for each one it is a constant character undergoing insignificant changes insofar as the plants are healthy and are raised in the same soil. In the experiments with this character, in order to obtain a confident difference, the long stem of 6–7 feet was united with the short one of ¾ to 1½ feet.

Each pair of the differing characters alluded to here were united through fertilisation.

For the

1st	experiment,	60	fertilisations	were	performed	on	15 plants.
2nd	"	58	"	"	"	"	10 "
3rd	"	35	"	"	"	"	10 "
4th	"	40	"	"	"	"	10 "
5th	"	23	"	"	"	"	5 "
6th	"	34	"	"	"	"	10 "
7th	"	37	"	"	"	"	10 "

Of a larger number of plants of the same kind, only the most vigorous were selected for fertilisation. Feeble specimens always yield uncertain results, because even in the first generation of the hybrids, and even more so in the following generations, some of the offspring either do not succeed in flowering or only produce few and inferior seeds.

Further, in all experiments reciprocal crosses were undertaken in this manner: one of the two kinds that served as seed plants for a number of fertilisations was used as the pollen plant for the other.

The plants were raised in garden beds, a small number of them in pots, and were kept in the natural upright position by means of poles, tree branches, and taut cords. For each experiment a number of potted plants were placed in a glasshouse during the flowering period. They served as a control for the main garden experiment in case of possible disturbance by insects. Among those that visit the pea plant, the beetle species Bruchus pisi could be dangerous for the experiment if it appears in large numbers. The female of this species is known to lay her eggs in the flowers and in doing so opens the keel; on the tarsi of one specimen caught in a flower, several pollen grains were obviously noticeable through a hand loupe. Here another circumstance must be noted in passing that could possibly give rise to the introduction of pollen from another individual. In rare individual cases certain parts of the otherwise completely normally developed flowers atrophy, which causes a partial exposure of the organs of fructification. Imperfect development of the keel was observed in which the style and anthers remained partially uncovered. It also sometimes happens that the pollen does not completely develop. In this case, a gradual lengthening of the style occurs during flowering until the stigma appears from the extremity of the keel. This curious phenomenon has been observed in hybrids of Phaseolus and Lathyrus.

The risk of adulteration by pollen from another individual is very slight for Pisum and can in no way disturb the result as a whole. With more than 10,000 carefully examined plants, the case of such undoubted interference occurred only a few times. Because no such disturbance was observed in the glasshouse, it may likely be supposed that Bruchus pisi and perhaps the previously alluded-to abnormalities in the flower structures are to blame.

The Form of the Hybrids

The experiments conducted with ornamental plants in past years already produced evidence that hybrids, as a rule, do not represent the precise intermediate form between the original parents. With individual characters that are particularly noticeable, like those related to the form and size of the leaves, to the pubescence of the individual parts, the intermediate form is in fact almost always apparent; in other cases, however, one of the two original parental characters possesses such an overwhelming dominance that it is difficult or quite impossible to find the other in the hybrid.

Such is exactly the behaviour of the Pisum hybrids. Each of the seven hybrid characters resembles one of the two original parental characters either so perfectly that the other one escapes observation or is so like it that a confident distinction cannot be made. This circumstance is of great importance for the determination and classification of the forms appearing among the progeny of the hybrids. In the following discussion those characters that are transmitted wholly or nearly unchanged in the hybrid association, that themselves represent the hybrid characters, are defined as *dominant*, and those that become latent in the association as *recessive*. The term "recessive" was chosen because the so-named characters recede or completely disappear in the hybrid, but among the progeny thereof, as will be shown later, reappear unchanged.

Further, it has been shown through all the experiments that it is completely unimportant whether the dominant character belongs to the seed plant or to the pollen plant; the hybrid form remains exactly the same in both cases. This interesting phenomenon deserves special notice, according to Gärtner, with the remark that even the most skilful expert is not able to distinguish in a hybrid which of the two united species was the seed or the pollen plant.

Of the differentiating characters introduced into the experiments, the following are dominant:

1. The spherical or somewhat rounded seed form with or without shallow indentations.
2. The yellow colour of the seed albumen.
3. The grey, grey-brown, or leather-brown colour of the seed coat, in association with violet-red flowers and reddish spotting in the leaf axils.
4. The simple inflated form of the pod.
5. The green colour of the unripe pod, associated with the same colour in the stem, leaf veins, and sepals.
6. The placement of the flowers along the stem.
7. The length of the longer stem.

With respect to this last character, it must be remarked that the longer of the two parent stems is generally surpassed by that of the hybrid, which may be attributed to the great luxuriance that appears in all parts of the plant when stems of very different length are united in the hybrid. Thus, for example, in repeated experiments stems of 1 foot and 6 foot length without exception produced stems in hybrid union whose length varied between 7 and 7½ feet. *In hybrids the seed coat* is often more spotted, and the spots sometimes blend together into small bluish-violet splotches. The spotting often appears even when it is absent in the original parental characters.

The hybrid forms of the *seed shape* and *albumen* develop directly after artificial fertilisation simply through the action of the pollen from another individual. Thus they can be observed within the first experimental year, whereas all of the others appear only in the following year in the plants raised from the fertilised seeds.

The First Generation of the Hybrids

In this generation, *along with the dominant* characters, the *recessive* characters reappear in their full individuality and do so in the determinate and pronounced average ratio of 3:1, so that of every four plants from this generation, three produce the dominant and one the recessive character. This applies without exception for all characters included in the experiment. The angular, wrinkled shape of the seeds, the green colour of the albumen, the white colour of the seed coat and of the flowers, the constriction of the pods, the yellow colour of the unripe pods, of the stems, sepals, and leaf veins, the umbel-formed inflorescence, and the dwarfed stem appear in these previously alluded-to numerical relationships emerging again without any essential difference. *Transitional forms were observed in none of the experiments.*

Because the hybrids produced from reciprocal crosses acquired a wholly similar form and because no appreciable variation appeared in their further development, the results for each experiment could be combined. The ratios acquired for each pair of two differing characters are as follows:

1st Experiment: Shape of the seeds. From 253 hybrids, 7,324 seeds were obtained in the second experimental year. Of these seeds 5,474 were round or somewhat rounded, and 1,850 angular wrinkled. The resulting ratio is 2.96:1.

2nd Experiment: Colour of the albumen. 258 plants produced 8,023 seeds, 6,022 yellow and 2,001 green; the former relate to the latter in the ratio 3.01:1.

In these experiments one generally gets both types of seeds in each pod. For well-developed pods that on average included 6 to 9 seeds, it was often the case that all of the seeds were round (Experiment 1) or all yellow (Experiment 2); more than 5 angular or 5 green, however, were never observed in one pod. It does not seem to make any difference if the pod develops earlier or later on the hybrid plant, if it belongs to the main stem or to a lateral one. With a few plants only single seeds developed in the pods formed first and these then had exclusively one of the two characters;

in the pods that formed later, however, the ratio remained normal. As in the individual pods, the distribution of characters varied similarly among individual plants. The first 10 members from both experimental sets serve as an illustration:

	1st Experiment		*2nd Experiment*	
	Shape of the seeds.		*Colour of the albumen.*	
Plant	*round*	*angular*	*yellow*	*green*
1	45	12	25	11
2	27	8	32	7
3	24	7	14	5
4	19	10	70	27
5	32	11	24	13
6	26	6	20	6
7	88	24	32	13
8	22	10	44	9
9	28	6	50	14
10	25	7	44	18

As extremes in the distribution of the two seed characters observed in *one* plant, in the first experiment 43 seeds were round and only 2 angular, and in another 14 round and 15 angular. In the second experiment 32 seeds were yellow and only 1 green, but also in another 20 yellow and 19 green.

These two experiments are important for ascertaining the mean ratios because they produce especially meaningful averages with a smaller number of experimental plants. While counting the seeds, however, especially in the 2nd experiment, some attention is required because in some seeds of several plants the green colour of the albumen is less developed and at first can be easily overlooked. The cause of the partial disappearance of the green colour has no relation to the hybrid character of the plants, as that occurs in the original parent plant as well; in addition, this peculiarity is limited only to the individual and is not inherited by the progeny. This phenomenon has often been observed in luxuriant plants. Seeds damaged by insects during their development often vary in colour and shape, but with some practice in sorting, errors are easily prevented. It is almost superfluous to mention that the pods must remain on the plant until they have ripened completely and have dried, because only then is the shape and colour of the seeds completely developed.

3rd Experiment: Colour of the seed coat. Of 929 plants, 705 produced violet-red flowers and grey-brown seed coats; 224 had white flowers and white seed coats. This results in a ratio of 3.16:1.

4th Experiment: Shape of the pods. Of 1,181 plants, 882 had simply inflated, 299 constricted pods. Hence the ratio 2.95:1.

5th Experiment: Colour of the unripe pod. The number of experimental plants was 580, of which 428 had green and 152 yellow pods. Thus the ratio of the former to the latter is 2.82:1.

6th Experiment: Position of the flowers. Of 868 cases, the flowers were located along the stem 651 times and were terminal 207 times. That ratio is 3.14:1.

7th Experiment: Length of the stem. Of 1064 plants, 787 had long stems, 277 short ones. Hence a relative ratio of 2.84:1. In this experiment, the dwarf plants were carefully dug up and moved to separate beds. This precaution was necessary because they would have atrophied among their tall intertwining siblings. In their youngest stages they are already easily distinguished by their compact growth and their dark-green thick leaves.

If the results of all experiments are summarised, there is an average ratio between the number of forms with dominant and recessive characters of 2.98:1 or 3:1.

The dominant character can have a *double signification* here, namely that of the original parental character or that of the hybrid character. Which of these two significations occurs in each case can only be determined in the next generation. An original parental character must be transmitted unchanged to all progeny, whereas the hybrid character must follow the same behaviour as observed in the first generation.

The Second Generation of the Hybrids

Those forms that preserve the recessive character in the first generation do not vary in the second generation in relation to that character; they remain *constant* in their progeny.

This is not the case for those that possess the dominant character in the first generation. Of these *two thirds* yield progeny that carry the dominant and recessive character in the ratio 3:1, and thus show the same behaviour as the hybrid forms; only *one third* remains constant with the dominant character.

The individual experiments produced the following results:

1st Experiment: Of 565 plants raised from round seeds of the first generation, 193 produced only round seeds and thus remained constant in this character; 372, however, simultaneously produced round and angular seeds in the ratio 3:1. Thus the number of hybrid types relative to the number of constant types is 1.93:1.

2nd Experiment: Of 519 plants raised from seeds whose albumen in the first generation had the yellow colour; 166 produced exclusively yellow, 353, however, produced yellow and green seeds in the ratio 3:1. This resulted in division of hybrid and constant forms in the ratio 2.13:1.

For each of the following experiments, 100 plants were selected that retained the dominant character in the first generation, and to test its signification, 10 seeds from each were cultivated.

3rd Experiment: The progeny of 36 plants produced exclusively grey-brown seed coats; from 64 plants some with grey-brown and some with white seed coats were produced.

4th Experiment: The progeny of 29 plants had only simply inflated pods, of 71, however, some had inflated, and some constricted pods.

5th Experiment: The progeny of 40 plants had only green pods, from those of 60 plants some had green and some yellow pods.

6th Experiment: The progeny of 33 plants had only flowers located along the stem, of 67, however, some had flowers located along the stem, and some had terminal flowers.

7th Experiment: The progeny of 28 plants produced long stems, from 72 plants some had long stems and some short.

In each of these experiments a particular number of plants with the dominant character is constant. For determination of the ratio in which segregation takes place for the forms with the constantly permanent character, the first two experiments are of special importance because a larger number of plants could be compared. The ratios 1.93:1 and 2.13:1, taken together, result almost precisely in the average ratio 2:1. The 6th experiment has almost an identical result; in the others the ratio fluctuates more or less, as must be expected given the small number of 100 experimental plants. The 5th experiment, which showed the largest deviation, was repeated and then, instead of the ratio 60:40, produced the ratio 65:35. *The average ratio 2:1 consequently appears certain.* Thus it is proved that of each form possessing the dominant character in the first generation, two thirds carry the hybrid character, one third however remains constant with the dominant character.

The ratio 3:1, which results in the distribution of the dominant and recessive characters in the first generation, resolves *then for all experiments into the ratio 2:1:1*, if one simultaneously distinguishes the dominant character in its signification as a hybrid character and as an original parental character. Because the members of the first generation arise directly from the seeds of the hybrids, *it now becomes apparent that the hybrids from each pair of differing characters form seeds, of which one half again develops the hybrid form, whereas the other yields plants that remain constant and produce in equal parts the dominant and recessive character.*

The Subsequent Generations of the Hybrids

The ratios according to which the offspring of the hybrids develop and segregate in the first and second generations are valid, in all probability, for all subsequent generations. The first and second experiments have now been continued through six generations, the third and seventh through five, the fourth, fifth, sixth through four generations, although beginning from the third generation with a smaller number of plants, without any noticeable deviation. The progeny of the hybrids in each generation segregated into hybrid and constant forms according to the ratio 2:1:1.

If A represents one of the two constant characters, for example the dominant, a the recessive, and Aa the hybrid form in which the two are united, then the expression

$$A + 2Aa + a$$

shows the developmental series for the progeny of the hybrids of each pair of divergent characters.

The observations made by Gärtner, Kölreuter, and others that hybrids possess the tendency to revert to the original parent-species is confirmed by the experiments herein discussed. It can be shown that the number of hybrids descended from fertilisation significantly decreases from generation to generation, without completely disappearing however, when compared to the number of forms and their progeny that have become constant. If one assumes that on average all plants in all generations have equally high fertility, and if one considers further that every hybrid forms seeds of which half arise again as hybrids, whereas the other half becomes constant with both characters in equal parts, then the numeric ratios for the progeny in each generation can be shown by the following tabulation, where A and a again represent the two original characters and Aa the hybrid form. For the sake of brevity, assume that every plant in every generation forms only 4 seeds.

				given as ratio:				
Generation	*A*	*Aa*	*a*	*A*	:	*Aa*	:	*a*
1	1	2	1	1	:	2	:	1
2	6	4	6	3	:	2	:	3
3	28	8	28	7	:	2	:	7
4	120	16	120	15	:	2	:	15
5	496	32	496	31	:	2	:	31
n				2^n-1	:	2	:	2^n-1

In the tenth generation, for example, there is $2^n-1 = 1023$. Of every 2048 plants that arise from this generation, there are 1023 that are constant for the dominant character, 1023 with the recessive character, and only 2 hybrids.

The Progeny of the Hybrids in Which Several Differing Characters Are Combined

For the experiments just discussed, plants were used that differed in only one essential character. The next objective consisted of researching whether the developmental law found for each pair of differing characters was valid when several different characters are united in the hybrid through fertilisation.

As for the form of the hybrids in this case, the experiments agreed in showing that the hybrid more closely resembles the original parent plant that possesses the larger number of dominant characters. If, for example, the seed plant has a short stem, terminal white flowers, and simple inflated pods whereas the pollen plant has a long stem, violet-red flowers along the stem, and constricted pods, then the hybrid reflects the seed plant only in the form of the pod; in the other characters it is identical to the pollen plant. If one of the two original parents possesses only dominant characters, then the hybrid is hardly or not at all distinguishable from it.

Two experiments were carried out with a larger number of plants. In the first experiment the original parent plants differed in the shape of the seeds and in the colour of the albumen; in the second in the shape of the seeds, the colour of the albumen, and in the colour of the seed coat. Experiments with seed characters lead to the simplest and most certain results.

To give an easier overview of these experiments, the differing characters of the seed plant are designated with *A*, *B*, *C*, those of the pollen plant with *a*, *b*, *c*, and the hybrid forms of these characters with *Aa*, *Bb*, *Cc*.

First Experiment:	*AB* seed plant,	*ab* pollen plant
	A round shape	*a* angular shape
	B yellow albumen	*b* green albumen

The seeds derived from fertilisation appeared round and yellow, resembling those of the seed plant. The plants raised from them produced seeds of four kinds that were often together in one pod. In total 556 seeds were produced from 15 plants; of these there were:

315 round and yellow,
101 angular and yellow,
108 round and green
32 angular and green.

All of them were cultivated the next year. Of the round yellow seeds 11 did not produce plants and 3 plants did not produce seeds. Among the remaining plants, there were:

38 round yellow seeds .... *AB*
65 round yellow and green seeds .... *ABb*
60 round yellow and angular yellow seeds .... *AaBb*
138 round yellow and green, angular yellow and green seeds ...... *AaBb*
From the angular yellow seeds 96 plants produced seeds, of which
28 had only angular yellow seeds .... *aB*
68 angular, yellow and green seeds .... *aBb*
From 108 round green seeds, 102 produced fruiting plants, from which there were:
35 with only round green seeds .... *Ab*
67 with round and angular green seeds .... *Aab*
The angular green seeds produced 30 plants with exactly this same type of seeds; they remained constant .... *ab*

The progeny of the hybrids thus appeared in 9 different forms and some in greatly unequal numbers. The following is the result when these are grouped and arranged:

38	plants designated	*AB*.
35	" "	*Ab*.
28	" "	*aB*.
30	" "	*ab*.
65	" "	*ABb*.
68	" "	*aBb*.
60	" "	*AaB*.

67	"	"	*Aab*.
138	"	"	*AaBb*.

All of the forms can be brought into 3 essentially different divisions. The first includes those with the designations *AB*, *Ab*, *aB*, *ab*; they possess only constant characters and vary no more in subsequent generations. Each of these forms is represented, on average, 33 times. The second group includes the forms *ABb*, *aBb*, *AaB*, *Aab*; these are constant in one character, hybrid in the other, and vary in the next generation only for the hybrid character. Each of them appears on average 65 times. The *AaBb* form appears 138 times, is hybrid for both characters, and behaves exactly like the hybrid from which it is derived.

If one compares the number of forms that occur in each of these divisions, the average ratios 1:2:4 are unmistakable. The numbers 33, 65, 138 are very close approximations of the ratio-numbers 33, 66, 132.

The developmental series thus consists of 9 classes. Four of them appear only one time each and are constant for both characters; the forms *AB*, *ab*, resemble the original parents; the other two represent the remaining possible constant combinations between the unions of characters *A*, *a*, *B*, *b*. Four classes appear twice each and are constant for one character, hybrid for the other. One class occurs four times and is hybrid for both characters. Thus the progeny of the hybrids, when two pairs of differing characters are combined in them, develop according to these terms:

$$AB + Ab + aB + ab + 2ABb + 2aBb + 2AaB + 2Aab + 4AaBb$$

This developmental series is indisputably a combination series in which the two developmental series for the characters *A* and *a*, *B*, and *b* are associated term by term. The total number of classes in the series is produced through combining the terms:

$$A + 2Aa + a$$
$$B + 2Bb + b$$

Second Experiment:	*ABC* seed plant,	*abc* pollen plant,
	A round form	*a* angular form
	B yellow albumen	*b* green albumen
	C grey-brown seed coat	*c* white seed coat

This experiment was conducted in a manner quite similar to the previous one. Of all the experiments it required the most time and effort. A total of 687 seeds were produced from 24 hybrids, all of which were spotted, coloured grey-brown or grey-green, round or angular. Of those, 639 plants produced seeds the following year and, as further researches showed, among them there were:

8	plants	*ABC*	22	plants	*ABCc*	45	plants	*ABbCc*
14	"	*ABc*	17	"	*AbCc*	36	"	*aBbCc*
9	"	*AbC*	25	"	*aBCc*	38	"	*AaBCc*
11	"	*Abc*	20	"	*abCc*	40	"	*AabCc*
8	"	*aBC*	15	"	*ABbC*	49	"	*AaBbC*
10	"	*aBc*	18	"	*ABbc*	48	"	*AaBbc*
10	"	*abC*	19	"	*aBbC*			
7	"	*abc*	24	"	*aBbc*			
			14	"	*AaBC*	78	"	*AaBbCc*
			18	"	*AaBc*			
			20	"	*AabC*			
			16	"	*Aabc*			

The developmental series includes 27 classes. Of those 8 are constant for all characters and each appears on average 10 times; 12 are constant for two characters, hybrid for the third, each appearing on average 19 times; 6 are constant for one character, hybrid for the other two, each of them occurring on average 43 times; one form appears 78 times and is hybrid for all characters. The ratio 10:19:43:78 appears so near to the ratio 10:20:40:80 or 1:2:4:8 that the latter without doubt represents the true values.

The development of the hybrids, when their original parents are different in 3 characters, results thusly according to the terms:

$$ABC + ABc + AbC + Abc + aBC + aBc + abC + abc + 2ABCc + 2AbCc + 2aBCc + 2abCc + 2ABbC + 2ABbc + 2aBbC + 2aBbc + 2AaBC + 2AaBc + 2AabC + 2Aabc + 2ABbc + 4ABbCc + 4aBbCc + 4AaBCc + 4AabCc + 4AaBbC + 4AaBbc + 8AaBbCc$$

Here too is a combination series in which the developmental series for the characters *A* and *a*, *B* and *b*, *C* and *c* are associated with each other. The terms:

$$A + 2Aa + a$$
$$B + 2Bb + b$$
$$C + 2Cc + c$$

reflect all classes in the series. The constant combinations that occur therein correspond to all combinations that are possible between the characters *A*, *B*, *C*, *a*, *b*, *c*; two of them, *ABC* and *abc*, resemble the two original parental plants.

In addition, various other experiments were undertaken with a smaller number of experimental plants in which the rest of the characters were associated in twos and threes in the hybrids; all produced approximately the same results. There is, then, no doubt that for all of the characters admitted into the experiments the following sentence is valid: *the progeny of hybrids in which several essentially differing characters are united represent the terms of a combination series in which the developmental series for each pair of differing characters are combined.* Simultaneously it thus is shown that *the behaviour of each pair of differing characters in hybrid association is independent of other differences between the two original parental plants.*

If n represents the number of the characteristic differences in the two original parent plants, then 3^n yields the number of classes in the combination series, 4^n the number of individuals that belong to the series, and 2^n the number of combinations that remain constant. Thus, for example, if the original parents have 4 different characters, the series includes $3^4 = 81$ classes, $4^4 = 256$ individuals, and $2^4 = 16$ constant forms; or in other words, among 256 progeny from hybrids there are 81 different combinations, of which 16 are constant.

All constant combinations that are possible in Pisum through combining the 7 characters previously alluded to were actually produced through repeated crosses. Their number is given as $2^7 = 128$. Simultaneously, factual evidence is produced *that constant characters occurring in different forms of a plant genus can, through repeated artificial fertilisation, occur in all possible combinations according to the rules of combination.*

Experiments on the flowering time of the hybrids are not yet concluded. It is already possible to note, in that regard, that flowering takes place at a time almost exactly intermediate between that of the seed and pollen plant, and the development of the hybrids in relation to this character will probably follow in the same manner as for the other characters. The forms chosen for experiments of this kind must differ in the mean flowering time by at least 20 days; further, it is necessary that the seeds when cultivated are placed at the same depth in the earth to produce simultaneous germination; and further, that large fluctuations in temperature during the entire flowering time must be taken into account to explain the resulting partial acceleration or retardation of flowering. Obviously this experiment has several difficulties that must be overcome, requiring great attention.

If we endeavour to summarise the results, we find that for those differing characters that admit easy and certain differentiation of the experimental plants, *we observe completely identical behaviour* in hybrid union. One half of the progeny of the hybrids for each pair of differing characters is also hybrid, whereas the other half is constant in equal proportions for the characters of the seed and pollen plants. If several differing characters are united in one hybrid through fertilisation, the progeny constitute the members of a combination series in which the developmental series for all pairs of differing characters are united.

The perfect identity shown by all characters tested in the experiment fully permits and justifies the assumption that the same behaviour applies to other characters that appear less sharply in the plants and thus could not be included in the individual experiments. An experiment with flower peduncles of differing lengths on the whole produced a rather satisfactory result, although distinction and classification of the forms could not be effected with the same certainty that is indispensable for correct experiments.

The Fertilising Cells of the Hybrids

The results of the initial experiments led to further experiments whose success appeared capable of throwing light on the nature of the germ and pollen cells of the hybrids. An important point of reference is offered in Pisum by the circumstance that constant forms appear in the progeny of its hybrids, and because they do so in all combinations of the united characters. Through experience, we find it to be invariably confirmed that constant progeny can be formed only when the germ cells and the fertilising pollen are the same, in that both are equipped with the ability to create perfectly equal individuals, as is the case with normal fertilisation of pure species. We must then treat it as necessary that the very same factors combine in the production of constant forms in the hybrid plant. Because the different constant forms are produced in *one* plant, even in *one* flower of the plant, it appears logical to assume that in the ovaries of the hybrids as many germ cells (germinal vesicles) and in the anthers as many pollen cells form as there are possible *constant* combination forms and that these germ and pollen cells correspond to the individual forms in their internal nature.

In fact, it can be shown theoretically that this assumption would be thoroughly ample to account for the development of the hybrids in individual generations, if one were simultaneously allowed to assume that the different kinds of germ and pollen cells are, on average, formed in equal numbers in the hybrid.

To test these assumptions experimentally, the following experiments were chosen: Two forms that differed constantly in the shape of the seeds and in the colour of the albumen were united through fertilisation. If the differing characters are once again represented as *A*, *B*, *a*, *b*, one has:

AB seed plant,	*ab* pollen plant
A round shape	*a* angular shape
B yellow albumen	*b* green albumen

The artificially fertilised seeds were cultivated along with seeds of the two original parent plants, and the most vigorous specimens were selected for reciprocal crosses. The fertilisations were:

1. The hybrid with the pollen of *AB*.
2. The hybrid " " " *ab*.
3. *AB* " " " the hybrid.
4. *ab* " " " the hybrid.

For each of these four experiments, all the flowers of three plants were fertilised. If the above assumption is true, then the germ and pollen cells must develop as forms *AB*, *Ab*, *aB*, *ab* in the hybrid; and united as:

1. The germ cells *AB*, *Ab*, *aB*, *ab* with the pollen cell *AB*.
2. " *AB*, *Ab*, *aB*, *ab* " *ab*.
3. " *AB* " *AB*, *Ab*, *aB*, *ab*.
4. " *ab* " *AB*, *Ab*, *aB*, *ab*.

From each of these experiments, then, only the following forms could emerge:

1. *AB*, *ABb*, *AaB*, *AaBb*.
2. *AaBb*, *Aab*, *aBb*, *ab*.
3. *AB*, *ABb*, *AaB*, *AaBb*.
4. *AaBb*, *Aab*, *aBb*, *ab*.

Further, if the individual forms of the germ and pollen cells of the hybrid were formed on average in equal numbers, then in each experiment the four previously stated combinations necessarily would be equal in their numerical relationships. A perfect accordance of the numerical ratios was not expected, however, because in every fertilisation, normal ones included, individual germ cells remain undeveloped or later atrophy, and even some of the well-developed seeds do not succeed in germinating after cultivation. Also, this assumption is limited in that the formation of the different germ and pollen cells merely approaches equality in numbers and not that every individual hybrid reaches such numbers with mathematical precision.

The *first and second* experiments had the main purpose of verifying the nature of the hybrid germ cells, and the *third and fourth* experiments of determining that of the pollen cells. As the above tabulation shows, the first and third experiments, and the second and fourth as well, should produce quite the same combinations; and, to some extent, these results should be partially apparent as early as the second year in the shape and colour of the artificially fertilised seeds. In the first and third experiments, the dominant characters of shape and colour, *A* and *B*, appear in each combination, one part in constant association, and the other part in hybrid union with the recessive characters *a* and *b*, and because of this they must impress their characteristic upon all of the seeds. All the seeds must therefore, if this assumption is true, appear round and yellow. In the second and fourth experiments, however, one combination is hybrid in shape and colour and the seeds are round and yellow; another is hybrid in shape and constant in the recessive character of colour and the seeds are round and green; the third is constant in the recessive character of shape and hybrid in colour and the seeds are angular and yellow; the fourth is constant in both recessive characters and the seeds are angular and green. In these two experiments, therefore, four seed types were expected, namely: round yellow, round green, angular yellow, and angular green.

The yield corresponds to these requirements perfectly.

There were obtained in the

1st experiment, 98 exclusively round yellow seeds;

3rd experiment, 94 " " " "

2nd experiment, 31 round yellow, 26 round green, 27 angular yellow, 26 angular green seeds;

4th experiment, 24 round yellow, 25 round green, 22 angular yellow, 27 angular green seeds.

There was no longer any doubt about a favourable result, but the next generation would produce the final determination. Of the cultivated seeds, in the following year, 90 plants in the first experiment, and 87 in the third, formed seeds. Of these, there were in the

First Experiment	*Third Experiment*	
20	25 round yellow seeds	*AB*
23	19 round yellow and green seeds	*ABb*
25	22 round and angular, yellow seeds	*AaB*
22	21 round and angular, yellow and green seeds	*AaBb*

In the second and fourth experiments the round and yellow seeds produced plants with round and angular, yellow and green seeds..........*AaBb*.

From the round green seeds, plants were produced with round and angular green seeds ....*Aab*.

The angular yellow seeds produced plants with angular yellow and green seeds .... *aBb*.

From the angular green seeds, plants were raised that produced once again only angular green seeds ....*ab*.

Although some seeds likewise did not germinate in these two experiments, the numbers found in the previous year could not be changed by that as each kind of seed produced plants that, in relation to their seeds, were similar among themselves and different from the others. Therefore, there were produced from the

Second Experiment	*Fourth Experiment*
31	24 plants with seeds of the form *AaBb*.
26	25 plants with seeds of the form *Aab*.
27	22 plants with seeds of the form *aBb*.
26	27 plants with seeds of the form *Aab*.

In all of the experiments, then, all forms appeared as this assumption required, and, in fact, in nearly the same numbers.

In another trial the characters of *flower colour and stem length* were admitted into the experiments and the selection designed so that in the third experimental year each character would appear in *half* of all plants if the above assumption were true. *A*, *B*, *a*, *b* serve once again as designations for the different characters.

A flowers violet-red	*a* flowers white.
B stem long	*b* stem short.

The form *Ab* was fertilised with *ab*, producing the hybrid *Aab*. Further, *aB* was also fertilised with *ab*, producing the hybrid *aBb*. In the second year the hybrid *Aab* was used as the seed plant for further fertilisation, and the other hybrid *aBb* as the pollen plant.

seed plant: *Aab*,	pollen plant: *aBb*
possible germ cells: *Ab*, *ab*,	pollen cells: *aB*, *ab*.

From the fertilisation between the possible germ and pollen cells, 4 combinations should appear, namely:

$$AaBb + aBb + Aab + ab$$

It thus becomes evident that according to the above assumption, in the third experimental year, of all plants

Half should have violet-red flowers	(*Aa*) . . .	groups:	1, 3
" should have white flowers	(*a*) . . .	"	2, 4
" should have a long stem	(*Bb*) . . .	"	1, 2
" should have a short stem	(*b*) . . .	"	3, 4

From 45 fertilisations in the second year, 187 seeds were produced, from which 166 plants succeeded in flowering in the third year. Of those, the individual groups appeared in the following numbers:

Group	*Flower Colour*	*Stem*	
1	violet-red	long	47 times
2	white	long	40 "
3	violet-red	short	38 "
4	white	short	41 "

Thus the violet-red flower colour	(*Aa*)	appeared in	85	plants
" white " "	(*a*)	" "	81	"
" long stem	(*Bb*)	" "	87	"
" short "	(*b*)	" "	79	"

The proposed theory finds ample confirmation in this experiment as well.

For the characters of the *pod form, pod colour, and flower placement*, smaller experiments were made and completely concurring results were produced. All combinations possible through the union of different characters appeared as expected and in nearly equal numbers.

Thus through experimental means the assumption is justified *that pea hybrids form germ and pollen cells that, according to their nature, correspond in equal numbers to all the constant forms that arise from the combination of characters united through fertilisation.*

The different forms among the progeny of the hybrids, as well as the numerical ratios in which they are observed, find a sufficient explanation in the principle just shown. The simplest case is offered by the developmental series for *each pair of differing characters*. It is known that this series is defined by the expression: $A + 2Aa + a$, in which A and a signify the forms with constant differing characters and Aa the hybrid form of both. It includes four individuals among the three different classes. In their formation, pollen and germ cells of the forms A and a occur in equal proportions on average in fertilisation, thus each form appears twice, since four individuals are formed. Therefore, participating in fertilisation are:

the pollen cells: $A + A + a + a$
the germ cells: $A + A + a + a$

It is a matter of chance which of the two kinds of pollen unites with each individual germ cell. According to the rules of probability, in the average of many cases, each pollen form A and a unites equally often with a germ cell form A and a; thus one of the two pollen cells A will converge with a germ cell A during the fertilisation, the other with a germ cell a and, in the same way, one pollen cell a will unite with a germ cell A, the other with a.

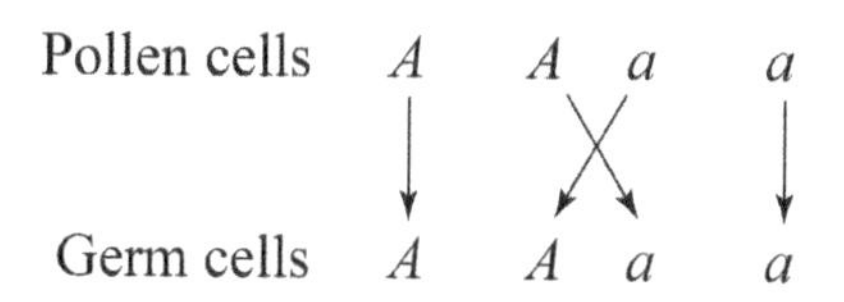

The result of fertilisations can be clearly illustrated if the designations for united germ and pollen cells are shown as fractions, with the pollen cells above the line, the germ cells below. Thus, in this case:

$$\frac{A}{A}+\frac{A}{a}+\frac{a}{A}+\frac{a}{a}$$

In the first and fourth classes the germ and pollen cells are the same, so the products of their association must be constant, A and a. With the second and third, however, once again a union of the two differing original parental characters takes place, hence the forms that appear from this fertilisation are completely identical to the hybrid from which they are derived. *Consequently, a repeated hybridisation takes place.* This accounts for the striking phenomenon that the hybrids are able, like the two original parental forms, to produce progeny that are identical to themselves; A/a and a/A both produce the same combination Aa, because, as alluded to earlier, it makes no difference for the result of fertilisation which of the two characters belongs to the pollen or germ cells. Thus,

$$\frac{A}{A}+\frac{A}{a}+\frac{a}{A}+\frac{a}{a}=A+2Aa+a$$

This is the *average* course for the self-fertilisation of hybrids when two differing characters are united in them. In individual flowers and in individual plants, the condition through which members of the series are formed, however, can undergo alterations that are not insignificant. Except for the fact that the numbers of both types of germ cells in the ovary can be supposed only on average to occur equally, it remains wholly left to chance which of the two kinds of pollen fertilises each individual germ cell. Thus the individual values necessarily undergo fluctuations and even extreme cases are possible as alluded to earlier in the experiments on seed shape and the colour of the albumen. The true numerical ratios can only be derived as the mean from the sum of the largest possible number of individual values; the larger their number, the more mere chance effects are eliminated.

The development series for hybrids in which *two kinds of differing characters* are associated includes 9 different forms with 16 individuals, namely: $AB + Ab + aB + ab + 2ABb + 2aBb + 2AaB + 2Aab + 4AaBb$. Between the different characters of the original parent plants A, a and B, b, 4 constant combinations are possible, thus the hybrid produces the corresponding 4 forms of germ and pollen cells: AB, Ab, aB, ab, and each of them will, on average, come into fertilisation 4 times, since 16 individuals are produced in the series. Thus, taking part in fertilisation are the

Pollen cells: $AB + AB + AB + AB + Ab + Ab + Ab + Ab + aB + aB + aB + aB + ab + ab + ab + ab$.

Germ cells: $AB + AB + AB + AB + Ab + Ab + Ab + Ab + aB + aB + aB + aB + ab + ab + ab + ab$.

In the average course of fertilisation, each pollen form unites equally often with every germ cell form, thus each of the 4 pollen cells AB unites once with each of the germ cell types AB, Ab, aB, ab. In precisely the same way, the union of the other pollen cells of forms Ab, aB, ab with all the other germ cells takes place. Consequently, one obtains

$$\frac{AB}{AB}+\frac{AB}{Ab}+\frac{AB}{aB}+\frac{AB}{ab}+\frac{Ab}{AB}+\frac{Ab}{Ab}+\frac{Ab}{aB}+\frac{Ab}{ab}+\frac{aB}{AB}+$$
$$\frac{aB}{Ab}+\frac{aB}{aB}+\frac{aB}{ab}+\frac{ab}{AB}+\frac{ab}{Ab}+\frac{ab}{aB}+\frac{ab}{ab}, \text{ or}$$

$AB + ABb + AaB + AaBb + ABb + Ab + AaBb + Aab + AaB + AaBb + aB + aBb + AaBb + Aab + aBb + ab = AB + Ab + aB + ab + 2ABb + 2aBb + 2AaB + 2Aab + 4AaBb$

The developmental series for hybrids can be accounted for in a quite similar manner when *three kinds of differing characters* are combined. The hybrid forms 8 different types of germ and pollen cells ABC, ABc, AbC, Abc, aBC, aBc, abC, abc and once again each pollen type unites on average once with each germ cell type.

The law of combination of the differing characters, by which the development of hybrids results, finds its *foundation and explanation* accordingly in the conclusive principle that hybrids produce germ and pollen cells corresponding in equal number to all constant forms that arise from the combination of the characters united through fertilisation.

Experiment on Hybrids of Other Plant Species

The objective for further experiments will be to ascertain whether the developmental law found for Pisum is also valid for hybrids of other plants. For this purpose several experiments were recently initiated. Two smaller experiments with species of Phaseolus have been concluded and they deserve mentioning here.

One experiment with Phaseolus vulgaris and Phaseolus nanus L. produced a completely corresponding result. Ph. nanus had, along with a dwarf stem, green, simply inflated pods; Ph. vulgaris, however, had a 10–12 foot long stem and yellow-coloured pods, that were constricted at the time of ripening. The numerical ratios in which the different forms occurred in the individual generations were the same as in Pisum. The development of constant combinations also resulted according to the law of simple combination of characters, precisely as is the case in Pisum. The results were:

Constant Combination:	*Stem:*	*Colour of the unripe pod:*	*Form of the ripe pod:*
1	long	green	inflated
2	"	"	constricted
3	"	yellow	inflated
4	"	"	constricted
5	short	green	inflated
6	"	"	constricted
7	"	yellow	inflated
8	"	"	constricted

Green pod colour, inflated form of the pod, and the long stem were dominant characters, as in Pisum.

Another experiment with two very different Phaseolus species had only partial success. Serving as the *seed plant* was Ph. nanus L., a very constant species with white flowers on short racemes and small white seeds in straight, inflated, and smooth pods; as the *pollen plant*, Ph. multiflorus W.

with a tall coiling stem, purple-red flowers on very long racemes, rough, sickle-shaped, bent pods and large seeds that are spotted and mottled black on a peach-flower-red background.

The hybrid was most similar to the pollen plant, only the flowers appeared less intensively coloured. Its fertility was very limited; of 17 plants that together produced many hundreds of flowers, a total of only 49 seeds were harvested. These were of intermediate size and retained markings resembling Ph. multiflorus; the background colour too was not essentially different. In the next year 44 plants were produced, of which only 31 succeeded in flowering. All characters of Ph. nanus that were latent in the hybrid reappeared in different combinations, but their ratios fluctuated greatly because of the small number of experimental plants; with individual characters, such as stem length and the pod form, the ratio was almost precisely 1:3, as in *Pisum*.

As limited as the result of this experiment may be for determining the numerical ratios in which the different forms occurred, it does, on the other hand, offer a case of a *curious transformation of colour* in the flowers and seeds of the hybrids. As is known to occur in Pisum, the characters of flower colour and seed colour appear unchanged in the first and subsequent generations, and the progeny of hybrids carry exclusively one or the other of the two original parental characters. Such behaviour is not the case in this experiment. The white flowers and seed colour of Ph. nanus, however, did appear the same in one rather fertile specimen in the first generation, but the other 30 plants developed flower colours that represent different gradations of purple-red to pale violet. The colour of the seed coat was no less different than that of the flower. No plant could be considered as perfectly fertile; some set no fruit at all, with others the pods developed only from the last flowers and never ripened. Well-formed seeds were harvested from only 15 plants. The greatest tendency toward sterility was shown in the forms with predominantly red flowers, in which only four ripe seeds were produced from 16 plants. Three of them had seed markings resembling Ph. multiflorus, but with a more or less pale background colour, the fourth plant produced only a single seed with a plain brown colour. The forms with prepotently violet-coloured flowers had dark-brown, black-brown, and completely black seeds.

The experiment was continued over two additional generations under similarly unfavourable conditions, as even among the progeny of rather fertile plants once again some were mostly less fertile or completely sterile. Flower and seed colours other than the ones noted did not appear. The forms that produced one or more of the recessive characters in the first generation remained constant for those characters without exception. Of those plants that acquired violet flowers and brown or black seeds, some displayed no change in flower and seed colour in the next generations; but the majority, in addition to completely similar progeny, produced some with white flowers and seed coats. The red-flowered plants remained so infertile that nothing in particular can be said about their further development.

Notwithstanding the many difficulties these observations had to confront, this experiment at least shows that the development of hybrids follows the same law as in *Pisum* in relation to those characters corresponding to the form of the plant. With respect to the colour characters, however, it seems difficult to find sufficient accordance. Laying aside the fact that a whole array of colours arises from the union of a white and purple-red colour, from purple to pale violet and white, it is a striking circumstance that of 31 plants that flowered, only one produced the recessive character of white colour, whereas with Pisum such is the case for every fourth plant on average.

But even these enigmatic phenomena might probably be explained according to the laws that are valid for Pisum if one could assume that the flower and seed colour of Ph. multiflorus are a complex of two or more completely independent colours that individually behave like other constant characters of a plant. If flower colour A were composed of the independent characters $A(1) + A(2) + \ldots$, that create the total impression of purple-red colour, then through fertilisation with the differing character of white colour a the hybrid combinations $A(1)a + A(2)a + \ldots$ would be formed, and similar behaviour would be expected with the corresponding colour of the seed coat. According to the assumption stated above, each of these hybrid colour combinations would be self-sufficient and would thus develop completely independently from the others. One can easily see, then, by combining the individual developmental series, a complete colour series must arise. If, for example, $A = A_1 + A_2$, then the hybrids A_1a and A_2a would correspond to the developmental series

$$A_1 + 2A_1a + a$$
$$A_2 + 2A_2a + a$$

The members of these series can occur in 9 different combinations and each of them represents the designation for another colour:

$1\ A_1A_2$	$2\ A_1a\ A_2$	$1\ A_2\ a$
$2\ A_1A_2a$	$4\ A_1a\ A_2a$	$2\ A_2a\ a$
$1\ A_1a$	$2\ A_1a\ a$	$1\ a\ a$

The numbers assumed for the individual combinations simultaneously indicate how many plants with the corresponding colour belong to the series. Since that sum equals 16, all colours, on average, are distributed to each of 16 plants, although, as the series itself shows, in unequal proportions.

If the development of colours actually took place in this manner, the case noted above could be explained—that white flowers and seed colour occurred only once among 31 plants of the first generation. This colour is included only once in the series and could thus, on average, develop once for each 16; and with three colour characters, only once for each 64 plants.

It must not be forgotten, however, that the explanation proposed here is based only on a mere supposition that has no other support than the very imperfect result of the experiment just discussed. It would, of course, be a worthwhile labour to follow the development of colour in hybrids with similar experiments, since it is probable that in this way we would come to understand the extraordinary multitude of *colours in our ornamental flowers.*

At this point, little more is known with certainty other than flower colour in most ornamental plants is an extremely variable character. The opinion has often been expressed that the stability of a species has been disrupted to a high degree or utterly broken through cultivation. There is a common inclination to refer to the development of cultivated forms as proceeding without rules and by chance; the colour of ornamental plants is generally cited as a pattern of instability. It is not apparent, however, why the mere placement in garden soil should result in such a drastic and persistent revolution in the plant organism. No one will seriously assert that the development of plants in a natural landscape is governed by different laws than in a garden bed. Here, just as there, typical variations must appear if the conditions of life are changed for a species, and it has the ability to adapt to the new conditions. It is freely admitted, through cultivation the production of new varieties is favoured, and by the hand of man many a variation is preserved that would have failed in the wild state, but nothing gives us the right to assume that the tendency for new varieties to form is so extremely augmented that species soon lose all stability and that their offspring break up into an infinite array of highly variable forms. If the change in the conditions of vegetation were the sole cause of variability, then one would be justified in expecting that those domesticated plants cultivated under almost the same conditions for centuries would have acquired stability. As is well known, this is not the case, for especially among them not only the most different but also the most variable forms are found. Only the Leguminosae, like Pisum, Phaseolus, Lens, whose organs of fructification are protected by a keel, constitute an appreciable exception. Even for these, numerous varieties have arisen during cultivation for more than 1000 years under the most diversified conditions; however, under the same permanent conditions of life, they retain stability similar to that of species growing in the wild.

It remains more than probable that there is a factor in action for the variability of cultivated plants, which hitherto has received little attention. Different experiences urge us to the view that our ornamental plants, with few exceptions, are *members of different hybrid series* whose legitimate further development is modified and delayed through numerous intercrosses. The circumstance must not be overlooked that cultivated plants usually are raised in larger numbers next to one another, which affords the most favourable opportunity for reciprocal fertilisation between the existing varieties and between species themselves. The probability of this view is corroborated by the fact that among the great host of variable forms, individuals are always found that remain constant in the one or the other character if every foreign influence is carefully prevented. These forms develop precisely the same as certain members of the complex hybrid series. Even with the most sensitive of all characters, that of colour, it cannot escape attentive observation that with individual forms the tendency toward variability occurs in very different degrees. Among plants that descend from *one* spontaneous fertilisation, there are often those whose progeny break up widely in the nature and arrangement of colours, whereas others produce forms with less distinction, and when a larger number individuals are examined, some are found that transmit flower colour unchanged to their progeny. The cultivated species of Dianthus give a demonstrative model. A white-flowering specimen of Dianthus Caryophyllus, that itself derives from a white-flowered variety, was isolated in a glasshouse during the flowering period; the numerous seeds acquired from it produced plants with absolutely the same white flower colour. A similar result was produced with a red-tinged-with-violet race crossed with a white-and-red-striped one. Many others, however, that were protected in the same manner, produced progeny with more or less different colours and markings.

Whoever surveys the colours in ornamental plants that arise from similar fertilisations, cannot easily avoid the conviction that here too development takes place according to a particular law that possibly can be expressed as the *combination of several independent colour characters.*

Concluding Remarks

It may not be without interest to compare the observations made herein on *Pisum* with the results of successful researches by the two authorities in this area, Kölreuter and Gärtner. According to their similar view, hybrids keep, in external appearance, either the form that is intermediate between the original parents, or they approach nearer to the type of the one or the other, sometimes hardly distinguishable from it. Generally, if fertilisation is effected through self-pollination, the seeds produced are of different forms that are distinct from the normal type. As a rule, the majority of the individuals from one fertilisation retain the hybrid form, whereas a few others become more similar to the seed plant and one or another individual appears to be nearer to the pollen plant. However, this does not apply to all hybrids without exception. For some individuals, the progeny more closely approach in part one, and in part the other, parental plant, or they all tend more to one side or the other; some, however, *remain perfectly similar to the hybrid* and continue unchanged. The hybrids of varieties behave like the hybrids of species, but they possess even greater variability in form and a more pronounced tendency to revert to the original parental forms.

In relation to the *attributes* of the hybrids and their resulting regular *development*, the agreement with observations made in Pisum is unmistakable. It is another matter for those cases mentioned as exceptions. Gärtner himself admits that the accurate determination of whether a form is more similar to one or the other of the two parents is often extremely difficult as it depends greatly on the subjective view of the observer. There could, however, be another circumstance that may have contributed to fluctuating and uncertain results in spite of the most careful observation and comparison. Plants that, for the most part, are considered as good species and that are different in a larger number of characters were used for the experiments. When dealing in general with cases of greater or lesser similarity, in addition to those characters that are clearly apparent, those that often are difficult to conceive in words must also be taken into account because, as every plant connoisseur knows, they are nevertheless sufficient to give the forms a strange appearance. Supposing that the development of hybrids takes place according to the laws applicable to Pisum, then the series of each single experiment must include very many forms, since it is well known that the number of terms increases by powers of 3 relative to the number of differing characters. With a relatively small number of experimental plants, then, the result could be only approximately clear and in individual cases could deviate not insignificantly. If, for example, the two original parents differed in 7 characters and if 100 to 200 plants were raised from the seeds of their hybrids to assess the degree of relatedness among the progeny, we can easily see how uncertain the judgment must become, since for 7 different characters, the developmental series consists of 2187 differing forms that include 16,384 individuals. One or another relationship could be overrepresented depending on which forms come by chance in larger numbers into the hands of the observer.

Further, if *dominant* characters, that are simultaneously transmitted completely or nearly unchanged to the hybrid, appear among those that differ, then of the two original parents, the one that possesses the larger number of dominant characters will be more apparent among the members of the developmental series. In the experiments with *Pisum* alluded to earlier for three differing characters, the dominant characters all belonged to the seed plant. Although the members of the series in their internal nature tend toward both original parental plants equally, in this experiment the seed plant type attained such great predominance that of every 64 plants of the first generation, 54 of them completely resembled it or differed from it in only one character. Under the circumstances, one sees how risky it can be to draw inferences about the internal relatedness of hybrids from external similarities.

Gärtner mentions that in those cases where development was regular, it was not the two original parents that were themselves preserved among the progeny of the hybrids but only single individuals closely related to them. With very extensive developmental series it could not, in fact, transpire otherwise. For 7 differing characters, for example, among more than 16,000 progeny of the hybrids, the two original parent forms appear only once each. Consequently, it is not readily probable that the two would be produced among a small number of experimental plants; with some probability, however, one may count on the appearance of individual forms that are close to one of them in the series.

We encounter *an essential difference* with those hybrids that remain constant in their progeny and propagate in the same way as the pure species. According to Gärtner, among these are the *distinctly fertile* hybrids: Aquilegia atropurpurea-canadensis, Lavatera pseudolbia-thuringiaca, Geum urbano-rivale and some Dianthus-hybrids, and, according to Wichura, hybrids of willow species. This circumstance is especially important for the evolutionary history of plants because constant hybrids acquire the status of *new species*. The truth of this fact has been authenticated by the highest observers and cannot be doubted. Gärtner had the opportunity to follow Dianthus Armeria-deltoides through the 10th generation, as it regularly propagated itself in the garden.

With Pisum, experiments showed that hybrids form *different* germ and pollen cells and that herein lies the reason for the variability of their progeny. Likewise, with other hybrids whose progeny behave similarly, we may assume the same cause; however, for those that remain constant, the assumption seems admissible that their fertilising cells are all the same, and are identical to the foundational cell of the hybrid. According to the view of famous physiologists, in phanerogams, for the purpose of reproduction, one germ cell and one pollen cell unite into a single cell[2] that is able to develop into an independent organism through the uptake of matter and the formation of new cells. This development takes place according to a constant law that is founded in the material nature and arrangement of the elements, which succeeds in a viable union in the cell. If the reproductive cells are the same and if they accord to the foundational cell of the mother plant, then the development of the new individual will be governed by the same law that applies to the mother plant. If there is a successful union of a germ cell with a *dissimilar* pollen cell, we must assume that between the elements of both cells that determine their reciprocal differences, there is some sort of counterbalance. The intervening cell that arises becomes the foundation of the hybrid organism whose development necessarily follows another law than for the two original parents. If the balance is assumed to be complete in the sense that the hybrid embryo is formed from similar cells in which the differences *are completely and permanently connected* then it can be further concluded that the hybrid, like every other autonomous plant species, will remain constant in its progeny. The reproductive cells that are formed in the ovaries and the anthers are the same and are identical to the underlying intervening cell.

In relation to those hybrids whose progeny are *variable*, one might perhaps assume that there is an intervention between the differing elements of the germ and pollen cells so that the formation of a cell as the foundation of the hybrid becomes possible; however, the counterbalance of opposing elements is only temporary and does not extend beyond the life of the hybrid plant. Because no changes are perceptible in the general appearance of the plant throughout the vegetative period, we must further infer that the differing elements succeed in emerging from their compulsory association only during development of the reproductive cells. In the formation of these cells, all existing elements act in a completely free and uniform arrangement in which only the differing ones reciprocally segregate themselves. In this manner the production of as many germ and pollen cells would be allowed as there are combinations of formative elements.

This attempted ascription of the essential distinction of either a *permanent or temporary association* of the differing cell elements in the development of the hybrids can, of course, only be of value as a hypothesis for which a wide scope of interpretation is possible given the dearth of reliable data. Some justification for the stated view lies in the evidence given for *Pisum* that the behaviour of each pair of differing characters in hybrid union is independent of the other differences between the two original plants and, further, that the hybrid produces as many types of germ and pollen cells as there are possible constant combination forms. The distinctive characters of two plants can ultimately rest only on differences in the nature and grouping of the elements that are present in their foundational cells in living interaction.

The validity of the set of laws suggested for Pisum requires additional confirmation and thus a repetition of at least the more important experiments would be desirable, for instance the one concerning the nature of the hybrid fertilising cells. An individual observer can easily miss a difference that, even if it at first seems unimportant, can increase in importance in such a way that it may not be neglected for the total result. Whether variable hybrids of other plant species reveal completely identical behaviour must also be determined through experiments; although one might well suppose that for important points a fundamental difference cannot occur since the *unity* of the evolutionary plan of organic life is beyond question.

In conclusion, special mention is deserved for the experiments carried out by Kölreuter, Gärtner, and others on the *transformation of one species into another through artificial fertilisation*. Special importance was set on these experiments; Gärtner counts them among the "most difficult in the production of hybrids."

For one species *A* to be converted into another *B*, both were united through fertilisation and the hybrids produced were again fertilised with the pollen of *B*; then the form was selected from the different offspring that was nearest to the species *B* and repeatedly fertilised with it and so on until one finally achieved a form that closely resembled *B* and remained constant in its progeny. Thus, the species *A* was transformed into the other species *B*. Gärtner himself conducted 30 such experiments with plants from the genera Aquilegia, Dianthus, Geum, Lavatera, Lychnis, Malva, Nicotiana, and Oenothera. The duration for the transformation was not the same for all species. While 3 fertilisations were sufficient for some, this had to be repeated 5–6 times with others; also, for these same species fluctuations were observed in different experiments. Gärtner attributes this difference to the circumstance that "the typical vigour with which a species acts in reproduction for changing and modifying the maternal type is very different with different plants, and that as a consequence the time periods within which and the number of generations through which one species is transformed into the other also must be different, so the transformation of some species is achieved through more, and of others through fewer, generations." Furthermore, the same observer noticed "that it also depends on the transformation processes which type and which individual is chosen for further transformation."

If one could assume that the development of the forms in these experiments took place in a manner similar to that in Pisum, then the whole transformation process could be explained rather simply. The hybrid forms as many kinds of germ cells as admissible given the constant combinations of its aggregated characters, and one of these is always the same as the fertilising pollen cells. Thus it is possible in all such experiments that as early as the second fertilisation a constant form resembling the pollen plant is acquired. Whether this is actually produced, however, depends in each individual case on the number of experimental plants as well as on the number of the differing characters that are united through fertilisation. Let us assume, for example, that the particular plants for the experiment were different in 3 characters and the species *ABC* was to be transformed into the other *abc* through repeated fertilisation with the pollen of that species. The hybrid arising from the first fertilisation forms 8 different kinds of germ cells, namely:

$$ABC,\ ABc,\ AbC,\ aBC,\ Abc,\ aBc,\ abC,\ abc.$$

These are again combined with the pollen cells *abc* in the second experimental year and one obtains the series:

$$AaBbCc + AaBbc + AabCc + aBbCc + Aabc + aBbc + abCc + abc$$

Because the form *abc* occurs once in the 8-membered series, it is less probable that it would be absent among the experimental plants, even if they were raised in a smaller number, and the transformation would be completed after just two fertilisations. If by chance it were not produced, then the fertilisation would need to be repeated on one of the nearest related combinations *Aabc, aBbc, abCc*. It becomes apparent that *the smaller the number of experimental plants and the larger the number of differing characters* in the two original parents, the longer such an experiment would have to be drawn out, and further that a postponement of one, or even of two, generations could easily occur with those same species, as Gärtner observed. The transformation of widely divergent species may well be finished only in the 5th or 6th experimental year because the number of different germ cells that are formed in the hybrid increases with the number of differing characters by powers of 2.

Gärtner found through repeated experiments that the *reciprocal* transformation time for some species is different so that often species *A* can be converted into another *B* one generation earlier than species *B* into the other *A*. He also derives from that evidence that the view of Kölreuter is not completely valid, according to which "the two natures in the hybrid are in perfect balance." It seems, however, that Kölreuter does not deserve such a criticism, and that, more importantly, Gärtner overlooked an important factor to which he himself draws attention elsewhere, that it is, namely, "dependent on which individual is chosen for further transformation." Experiments in this regard made with two Pisum species indicate that which species is being transformed into the other can make a great difference for the selection of the most suitable individuals for the purpose of further fertilisation. The two experimental plants differed in 5 characters, and species *A* possessed all dominant, the other *B* all recessive characters. For the reciprocal transformation *A* was fertilised with the pollen from *B* and conversely *B* with that of *A*, then the same was repeated with both hybrids in the next year. With the first experiment *B/A* there were in the third experimental year 87 plants, in fact *in all possible 32 forms*, available for selection of individuals for further fertilisation; for the second experiment *A/B* 73 plants were produced that in their general appearance were thoroughly *identical to the pollen plant*, but according to their internal nature were necessarily as different as the forms of the other experiment. Calculated selection was then only possible in the first experiment, in the second some plants had to be rejected by mere chance. Of the latter only a portion of the flowers were fertilised with the pollen of *A*, the others were left to self-fertilise. As the next year's cultivation showed, of every 5 plants used for fertilisation for the two experiments, there was in accordance with the pollen plant:

First Experiment	Second Experiment	
2 plants	----	in all characters
3 "	----	" 4 "
----	2 plants	" 3 "
----	2 "	" 2 "
----	1 plant	" 1 character.

For the first experiment, the transformation was finished; for the second, which was not continued further, two additional fertilisations would probably have been necessary.

Even though the case does not frequently occur in which the dominant characters belong exclusively to the one or the other original parent plant, it still makes a difference *which* of the two possesses the greater number. If the majority of the dominant characters belong to the pollen plant, then the selection of forms for further fertilisation will give a lower degree of certainty compared to the converse case, resulting in a lengthened time required for transformation, supposing that the experiment is viewed as finished only when a form is produced that not only appears the same as the pollen plant in its form but also likewise remains constant in its progeny.

Through the success of transformation experiments, Gärtner was persuaded to oppose the opinion of those naturalists who dispute the stability of plant species and assume continuous evolution of plant species. He sees in the completed transformation of one species into the other the unambiguous evidence that a species has fixed limits beyond which it cannot change. Although this view cannot be afforded unconditional validity, nonetheless a confirmation deserving notice regarding the supposition made earlier about the variability of cultivated plants is found in the experiments performed by Gärtner.

Among the experimental species were cultivated plants, like Aquilegia atropurpurea and canadensis, Dianthus Caryophyllus, chinensis and japonicus, Nicotina rustica and paniculata, and these too had lost none of their autonomy after 4 or 5 hybrid unions.

Notes

Preface

1. James L. Farmer and Daniel J. Fairbanks, "Interaction of the *bw* and *w* Loci in *Drosophila melanogaster*," *Genetics* 107 (1984): s30.

2. Evelyn F. Keller, *A Feeling for the Organism: The Life and Work of Barbara McClintock* (San Francisco: W. H. Freeman, 1983).

3. Lennard Bickel, *Facing Starvation: Norman Borlaug and the Fight against Hunger* (Pleasantville, NY: Reader's Digest Press, 1974).

4. Hugo Iltis, *Gregor Mendel: Leben, Werk und Wirkung* (Berlin: Springer, 1924).

5. Hugo Iltis, *Life of Mendel*, trans. Eden Paul and Cedar Paul (New York: Hafner, 1966).

6. Vítězslav Orel, *Gregor Mendel: The First Geneticist*, trans. Stephen Finn (Oxford: Oxford University Press, 1996).

7. Jan Klein and Norman Klein, *Solitude of a Humble Genius—Gregor Johann Mendel*, vol. 1, *Formative Years* (Berlin: Springer, 2013).

8. Robert C. Olby, *Origins of Mendelism*, 2nd ed. (Chicago: University of Chicago Press, 1985).

9. Curt Stern and Eva R. Sherwood, *The Origin of Genetics: A Mendel Source Book* (San Francisco: W. H. Freeman, 1966).

10. Orel, *Gregor Mendel*, 157–60.

Prologue

1. "Wetter vom 11 Jänner," *Mährischer Correspondent*, Brünn, January 13, 1865, 7, http://www.digitalniknihovna.cz/mzk/view/uuid:04530090-64ba-11e3-8c6a-005056825209?page=uuid:08aa19f0-6710-11e3-8387-001018b5eb5c.

2. Daniel J. Fairbanks and Scott Abbott, "Alexander Makowsky's January 1865 Lecture 'On Darwin's Theory of Organic Creation': An English Translation with Commentary," *Folia Mendeliana* 55 (2019): 11.

3. Fairbanks and Abbott, "Alexander Makowsky's January 1865 Lecture," 11–13.

4. Daniel J. Fairbanks, "Mendel and Darwin: Untangling a Persistent Enigma," *Heredity* 124 (2020): 265, https://www.nature.com/articles/s41437-019-0289-9.

5. Jan Klein and Norman Klein, *Solitude of a Humble Genius—Gregor Johann Mendel*, vol. 1, *Formative Years* (Berlin: Springer, 2013), 295.

6. Thomas H. Huxley, "Science and 'Church Policy,'" *Reader*, December 31, 1864.

7. Arthur D. Darbishire, *Breeding and the Mendelian Discovery* (London: Cassell and Company, 2011), 189.

8. Loren Eisley, *Darwin's Century: Evolution and the Men Who Discovered It* (London: Scientific Book Guild, 1959), 206.

9. Robert C. Olby, *Origins of Mendelism*, 2nd ed. (Chicago: University of Chicago Press, 1985), 220.

10. Olby, *Origins of Mendelism*, 221.

11. English translation by Daniel J. Fairbanks from "Monats-Versammlung des naturforschenden Vereins in Brünn am 8 März 1865," *Brünner Zeitung*, no. 65, March 20, 1865, http://www.digitalniknihovna.cz/mzk/view/uuid:e32cb9e0-f061-11e3-a012-005056825209.

12. Curt Stern and Eva R. Sherwood, *The Origins of Genetics: A Mendel Source Book* (San Francisco: W. H. Freeman, 1966), 60–61.

13. Christies of London, "Important Scientific Books from the Collection of Peter and Margarethe Braune, sale 17700, lot 307," https://www.christies.com/lotfinder/Lot/mendel-johann-gregor-1822-1884-versuche-uber-pflanzen-hybriden-6216731-details.aspx.

14. Charles Darwin, *On the Origin of Species by Means of Natural Selection, or the Preservation of Favoured Races in the Struggle for Life*, 1st ed. (London: John Murray, 1859), 13.

15. Fairbanks, "Mendel and Darwin," 269.

16. Allan Franklin, A. W. F. Edwards, Daniel J. Fairbanks, Daniel L. Hartl, and Teddy Seidenfeld, *Ending the Mendel–Fisher Controversy* (Pittsburgh, PA: University of Pittsburgh Press, 2008).

17. Zhores A. Medvedev, *The Rise and Fall of T. D. Lysenko*, trans. I. Michael Lerner (New York: Columbia University Press, 1969), 58.

Chapter 1

1. Ethnic Germans and Czechs lived in most of the cities and towns associated with Mendel's life and carried both German and Czech names at the time. These cities and towns are now within the Czech Republic and carry only the Czech

names. As an ethnic German, Mendel knew them mostly by their German names. For this reason, I use the German name of each city at first mention (unless the city has a common English name, such as Vienna or Prague), followed by the Czech name in parentheses and the German name thereafter. The exception is in the epilogue, where I used Czech names because its setting is the twentieth and twenty-first centuries.

2. A detailed account of the various arguments for July 22 as opposed to July 20, 1822, as Mendel's birth date is in Jan Klein and Norman Klein, *Solitude of a Humble Genius—Gregor Johann Mendel*, vol. 1, *Formative Years* (Berlin: Springer, 2013), 121–25; Vítězslav Orel, *Gregor Mendel: The First Geneticist*, trans. Stephen Finn (Oxford: Oxford University Press, 1996), 36; and Peter van der Pas, "The Date of Gregor Mendel's Birth," *Folia Mendeliana* 7 (1972): 7–12.

3. Klein and Klein, *Solitude of a Humble Genius*, 125.

4. Klein and Klein, *Solitude of a Humble Genius*, 125.

5. Hugo Iltis, *Life of Mendel*, trans. Eden Paul and Cedar Paul (New York: Hafner, 1966), 34.

6. Hugo Iltis, *Life of Mendel*, 34.

7. Anne Iltis, "Gregor Mendel's Autobiography," *Journal of Heredity* 45 (1954): 234.

8. Iltis, "Gregor Mendel's Autobiography," 234.

9. Iltis, *Life of Mendel*, 35.

10. An English translation of these poems by Eden Paul and Cedar Paul is in Hugo Iltis, *Life of Mendel*. However, the translation has several flaws, so the translation provided here is a new one by Scott Abbott and Daniel J. Fairbanks, published for the first time here, from the original German in Hugo Iltis, *Gregor Mendel: Leben, Werk und Wirkung*, 14–15, https://www.google.com/books/edition/Gregor_Johann_Mendel/AD0MAAAAMAAJ.

11. Iltis, "Gregor Mendel's Autobiography," 234.

12. Iltis, *Life of Mendel*, 39.

13. This serendipitous timing was pointed out by Klein and Klein, *Solitude of a Humble Genius*, 185.

14. Klein and Klein, *Solitude of a Humble Genius*, 185.

Chapter 2

1. Anne Iltis, "Gregor Mendel's Autobiography," *Journal of Heredity* 45 (1954): 231–34.

2. Iltis, "Gregor Mendel's Autobiography," 234.

3. Jan Klein and Norman Klein, *Solitude of a Humble Genius—Gregor Johann Mendel*, vol. 1, *Formative Years* (Berlin: Springer, 2013), 186.

4. Peter van Dijk and T. H. Noel Ellis, "Mendel's Journey to Paris and London: Context and Significance for the Origin of Genetics," *Folia Mendeliana* 56

(2020): 5–34; Peter van Dijk, "Gregor Mendel's Meeting with Pope Pius IX: The Truth in the Story," *Folia Mendeliana* 56 (2020): 35–50.

5. Klein and Klein, *Solitude of a Humble Genius*, 230.

6. Anna Matalová, "A Monument to F. M. Klácel (1809–1882) in the Vicinity of the Mendel Statue in Brno," *Folia Mendeliana* 14 (1973): 252.

7. Margaret H. Peaslee and Vítězslav Orel, "The Evolutionary Ideas of F. M. (Ladimir) Klacel, Teacher of Gregor Mendel," *Biomedical Papers of the Medical Faculty of the University of Palacky, Olomouc, Czech Republic* 151 (1972): 152.

8. Matalová, "A Monument to F. M. Klácel," 251–63.

9. Klein and Klein, *Solitude of a Humble Genius*, 230.

10. Klein and Klein, *Solitude of a Humble Genius*, 230.

11. Vítězslav Orel, *Gregor Mendel: The First Geneticist*, trans. Stephen Finn (Oxford: Oxford University Press, 1996).

12. Klein and Klein, *Solitude of a Humble Genius*, 234.

13. John Tyrrell, *Janáček: Years of a Life—Volume 1 (1854–1914): The Lonely Blackbird* (London: Faber and Faber, 2006).

14. Orel, *Gregor Mendel*, 54.

15. Vítězslav Orel and Antonín Verbík, "Mendel's Involvement in the Plea for Freedom of Teaching in the Revolutionary Year of 1848," *Folia Mendeliana* 19 (1984): 223–33.

16. Klein and Klein, *Solitude of a Humble Genius*, 280.

17. Klein and Klein, *Solitude of a Humble Genius*, 281.

18. Klein and Klein, *Solitude of a Humble Genius*, claim on p. 281 that the handwriting is "'unequivocally' Mendel."

19. Klein and Klein, *Solitude of a Humble Genius*, 253–54.

20. Hugo Iltis, *Life of Mendel*, trans. Eden Paul and Cedar Paul (New York: Hafner, 1966), 56.

21. Klein and Klein, *Solitude of a Humble Genius*, 254.

22. Iltis, *Life of Mendel*, 54.

23. Iltis, *Life of Mendel*, 58.

24. Original English translation from Daniel J. Fairbanks, "Mendel and Darwin: Untangling a Persistent Enigma," *Heredity* 124 (2020): 264, https://www.nature.com/articles/s41437-019-0289-9, from a German transcription in Vítězslav Orel, Gerhard Czihak, and Hans Wieseneder, "Mendel's Examination Paper on the Geological Formation of the Earth of 1850," *Folia Mendeliana* 18 (1983): 237.

25. Iltis, *Life of Mendel*, 71–72.

26. Iltis, *Life of Mendel*, 72.

27. Orel, *Gregor Mendel*, 66.

28. Orel, *Gregor Mendel*, 66; see also Iltis, *Life of Mendel*, 76.

29. Iltis, *Life of Mendel*, 76.

Chapter 3

1. Hugo Iltis, *Life of Mendel*, trans. Eden Paul and Cedar Paul (New York: Hafner, 1966), 76.
2. Iltis, *Life of Mendel*, 77.
3. Johann Vollmann and Anna Matalová, "Echoes of Mendel's Life and Work in Newspapers between the Years 1850–1884," *Folia Mendeliana* 52, no. 1 (2016): 22.
4. Sander Gliboff, "Evolution, Revolution, and Reform in Vienna: Franz Unger's Ideas on Descent and Their Post-1848 Reception," *History of Science* 37 (1998): 223.
5. Franz Unger, *Botanical Letters to a Friend*, trans. Eden Paul (Philadelphia: Lindsay and Blakiston, 1853), 107, https://www.biodiversitylibrary.org/item/215881#page/15/mode/1up.
6. Vítězslav Orel, "Mendel and New Scientific Ideas at the Vienna University," *Folia Mendeliana* 7 (1972): 31–32.
7. Franz Unger, *Botanische Briefe* (Vienna: Verlag von Carl Gerold & Sohn, 1852), https://www.biodiversitylibrary.org/item/47372#page/11/mode/1up.
8. Unger, *Botanical Letters to a Friend*.
9. Unger, *Botanical Letters to a Friend*, 92.
10. Gliboff, "Evolution, Revolution, and Reform in Vienna," 199.
11. Unger, *Botanical Letters to a Friend*, 116.
12. Robert C. Olby, *Origins of Mendelism*, 2nd ed. (Chicago: University of Chicago Press, 1985), 201.
13. Gliboff, "Evolution, Revolution, and Reform in Vienna," 201.
14. Franz Unger, *Die Urwelt in ihren verschiedenen Bildungsperioden* (Vienna: Fr. Beck, 1851), https://gdz.sub.uni-goettingen.de/id/PPN782695469.
15. Unger, *Die Urwelt in ihren verschiedenen Bildungsperioden*, 39, English translation by Daniel J. Fairbanks.
16. Gliboff, "Evolution, Revolution, and Reform in Vienna," 195–96.
17. Olby, *Origins of Mendelism*, 202.
18. Scott Abbott and Daniel J. Fairbanks, "Experiments on Plant Hybrids by Gregor Mendel," *Genetics* 204 (2016): 420, https://www.genetics.org/content/204/2/407.
19. Jan Klein and Norman Klein, *Solitude of a Humble Genius—Gregor Johann Mendel*, vol. 1, *Formative Years* (Berlin: Springer, 2013), 349.
20. Daniel J. Fairbanks and Scott Abbott, "Darwin's Influence on Mendel: Evidence from a New Translation of Mendel's Paper," *Genetics* 204 (2016): 401–5, https://www.genetics.org/content/204/2/401.
21. Klein and Klein, *Solitude of a Humble Genius*, 295.
22. The original German is in Hugo Iltis, *Gregor Johann Mendel: Leben, Werk und Wirkung* (Berlin: Springer, 1924), 83, https://www.google.com/books/edition/Gregor_Johann_Mendel/AD0MAAAAMAAJ, English translation by Daniel J. Fairbanks.

23. Vollmann and Matalová, "Echoes of Mendel's Life and Work in Newspapers between the Years 1850–1884," 22–23.

24. Gregor Mendel, "Über Verwüstung am Gartenrettig durch Raupen," *Verhandlungen des zoologisch-botanischen Vereins in Wien* 3 (1853): 116–18, https://www.biodiversitylibrary.org/bibliography/16346.

25. Gregor Mendel, "Beschreibung des sogenannten Erbsenkäfers, *Bruchus pisi*, Mitgeteilt von V. Kollar," *Verhandlungen des zoologisch-botanischen Vereins in Wien* 4 (1854): 27–30, https://www.biodiversitylibrary.org/bibliography/16346.

Chapter 4

1. Vítězslav Orel, *Gregor Mendel: The First Geneticist*, trans. Stephen Finn (Oxford: Oxford University Press, 1996), 84.

2. Orel, *Gregor Mendel*, 85; Jan Klein and Norman Klein, *Solitude of a Humble Genius—Gregor Johann Mendel*, vol. 1, *Formative Years* (Berlin: Springer, 2013), 295.

3. Orel, *Gregor Mendel*, 85

4. Klein and Klein, *Solitude of a Humble Genius*, 296–97.

5. Hugo Iltis, *Life of Mendel*, trans. Eden Paul and Cedar Paul (New York: Hafner, 1966), 94.

6. Jaroslav Kříženecký, "Mendels zweite erfolglose Lehramtsprüfung im Jahre 1856," *Sudhoffs Archiv für Geschichte der Medizin und der Naturwissenschaften* 47 (1963): 308; Klein and Klein, *Solitude of a Humble Genius*, 364.

7. Nowotný's first name is uncertain. Hugo Iltis, who conducted the interview, refers to him only as "Inspektor Nowotny," without providing a first name (Hugo Iltis, *Gregor Mendel: Leben, Werk und Wirkung* [Berlin: Springer, 1924], 71). Klein and Klein, *Solitude of a Humble Genius*, presumed that his name was Adolph Nowotny. However, I found in the list of members of the Natural Science Society in Brünn (1869) "Johann Nowotný," listed with the title "Lehrer an der Normalhauptschule in Brünn" ("teacher in the normal high school in Brünn"), as the only member with that surname in the list. Given that the Nowotny listed by Iltis was a teacher at the same time as Mendel, it is likely that the Johann Nowotný and Inspektor Nowotny are the same person.

8. Klein and Klien, *Solitude of a Humble Genius*, 363–64.

9. Iltis, *Life of Mendel*, 95.

10. Orel, *Gregor Mendel*, 81.

11. Iltis, *Life of Mendel*, 89.

12. Max Toperczer, "Liznar, Josef (1852–1932), Geophysiker und Meteorologe," *Österreichishes Biographisches Lexikon* 5 (1971): 254, https://www.biographien.ac.at/oebl/oebl_L/Liznar_Josef_1852_1932.xml.

13. Iltis, *Life of Mendel*, 89–90.

14. Orel, *Gregor Mendel*, 81.

15. Josef Auspitz to Gregor Mendel, Brünn, March 13, 1858. Collection of the Mendelianum of the Moravian Museum, Brno, Czech Republic. English translation by Scott Abbott and Daniel J. Fairbanks.

16. Iltis, *Life of Mendel*, 88–89.

17. Iltis, *Life of Mendel*, 88–89.

18. Iltis, *Life of Mendel*, 92.

19. Iltis, *Life of Mendel*, 105.

20. Robin M. Henig, *The Monk in the Garden: The Lost and Found Genius of Gregor Mendel, the Father of Genetics* (Boston: Houghton Mifflin Harcourt, 2000), 15–16.

21. For examples of authors who claimed Mendel conducted hybridization experiments with mice, see Siddhartha Mukherjee, *The Gene: An Intimate History* (New York: Scribner, 2016); Kenneth Paigen, "One Hundred Years of Mouse Genetics: An Intellectual History. I. The Classical Period (1902–1980)," *Genetics* 163 (2003): 1–7, https://www.genetics.org/content/163/1/1; Louise van der Weyden, Jaqueline K. White, David J. Adams, and Darren W. Logan, "The Mouse Genetic Toolkit: Revealing Function and Mechanism," *Genome Biology* 12 (2011): 224, https://genomebiology.biomedcentral.com/articles/10.1186/gb-2011-12-6-224; Megan Phifer-Rixey and Michael W. Nachman, "Insights into Mammalian Biology from the Wild House Mouse *Mus musculus*," *eLife* 4: e05959, https://elifesciences.org/articles/05959; and R. Allan Reese, "Darwin, Mendel, and the Evolution of Evolution," *Significance* 6 (2009): 127–29, https://rss.onlinelibrary.wiley.com/doi/full/10.1111/j.1740-9713.2009.00377.x.

22. John A. Weir, "Agassiz, Mendel, and Heredity," *Journal of the History of Biology* 1 (1968): 201.

23. Iltis, *Life of Mendel*, 95.

24. Scott Abbott and Daniel J. Fairbanks, "Experiments on Plant Hybrids by Gregor Mendel," *Genetics* 204 (2016): 407.

25. B. E. Bishop, "Mendel's Opposition to Evolution and to Darwin," *Journal of Heredity* 87 (1996): 212.

26. L. A. Callender, "Gregor Mendel: An Opponent of Descent with Modification," *History of Science* 26 (1988): 72.

27. Daniel J. Fairbanks and Bryce Rytting, "Mendelian Controversies: A Botanical and Historical Review," *American Journal of Botany* 88 (2001): 737–52, https://doi.org/10.2307/2657027.

28. Daniel J. Fairbanks and Scott Abbott, "Darwin's Influence on Mendel: Evidence from a New Translation of Mendel's Paper," *Genetics* 204 (2016): 401–5, https://www.genetics.org/content/204/2/401.

29. Herbert F. Roberts, *Plant Hybridization Before Mendel* (Princeton, NJ: Princeton University Press, 1929), https://www.biodiversitylibrary.org/item/23454#page/9/mode/1up.

30. Abbott and Fairbanks, "Experiments on Plant Hybrids by Gregor Mendel," 407.

31. Abbott and Fairbanks, "Experiments on Plant Hybrids by Gregor Mendel," 407.

32. Carl F. Gärtner, *Versuche und Beobachtungen über die Bastarderzeugung im Pflanzenreich* (Stuttgart: H. F. Herring, 1849).

33. Robert C. Olby, *Origins of Mendelism*, 2nd ed. (Chicago: University of Chicago Press, 1985), 213.

34. Jiří Sekerák, "Mendel's Scientific Society in Brno," *Folia Mendeliana* 54 (2018): 5–10.

35. The details of how the Natural Science Society was formed are recounted by Orel, *Gregor Mendel*, 89–91.

36. This summary of reports to the Natural Science Society of Brünn on plant hybridization is from Orel, *Gregor Mendel*, 128–29.

37. Abbott and Fairbanks. "Experiments on Plant Hybrids by Gregor Mendel," 407.

38. Charles Darwin, *On the Origin of Species by Means of Natural Selection, or the Preservation of Favoured Races in the Struggle for Life*, 3rd ed. (London: John Murray, 1861), 525.

39. A number of revisionist historians claim that Mendel never described the principles of segregation and independent assortment in his article. His statements of these principles, however, are readily evident, although he does not give them these titles. See Fairbanks and Rytting, "Mendelian Controversies," 737–52.

40. Robert C. Olby, "Mendel No Mendelian?," *History of Science* 17 (1979): 53–54.

41. Olby, *Origins of Mendelism*, 220–21; Hui Zhang, Wen Chen, and Kun Sun, "Mendelism: New Insights from Gregor Mendel's Lectures in Brno," *Genetics* 207 (2017): 1–8.

Chapter 5

1. Daniel L. Hartl and Daniel J. Fairbanks, "Mud Sticks: On the Alleged Falsification of Mendel's Data," *Genetics* 175 (2007): 975, https://www.genetics.org/content/175/3/975.

2. Curt Stern and Eva R. Sherwood, *The Origin of Genetics: A Mendel Source Book* (San Francisco: W. H. Freeman, 1966), 61.

3. An anonymous reporter, probably Alexander Makowsky, for the *Neuigkeiten*, a daily newspaper in Brünn, reported on the meeting in the February 9 issue, the first paragraph of the article stating, "At the monthly meeting held yesterday the newly-elected Vice-President, Herr [Carl] Theimer, took the chair. The meeting was very well attended. After the reading of the communications received, Herr Professor G. Mendel delivered a long lecture, of special interest to botanists, on plant hybrids raised by artificial fertilisation of related species, that is by transfer of the male pollen to the seed plant." This English translation is from Robert C. Olby, *Origins of Mendelism*, 2nd ed. (Chicago: University of Chicago Press, 1985), 220.

4. Scott Abbott and Daniel J. Fairbanks, "Experiments on Plant Hybrids by Gregor Mendel," *Genetics* 204 (2016): 407.
5. Abbott and Fairbanks, "Experiments on Plant Hybrids by Gregor Mendel," 407.
6. Abbott and Fairbanks, "Experiments on Plant Hybrids by Gregor Mendel," 408.
7. Abbott and Fairbanks, "Experiments on Plant Hybrids by Gregor Mendel," 408.
8. Abbott and Fairbanks, "Experiments on Plant Hybrids by Gregor Mendel," 410.
9. Abbott and Fairbanks, "Experiments on Plant Hybrids by Gregor Mendel," 412.
10. Abbott and Fairbanks, "Experiments on Plant Hybrids by Gregor Mendel," 409.
11. Abbott and Fairbanks, "Experiments on Plant Hybrids by Gregor Mendel," 409.
12. Abbott and Fairbanks, "Experiments on Plant Hybrids by Gregor Mendel," 410.
13. Abbott and Fairbanks, "Experiments on Plant Hybrids by Gregor Mendel," 410.
14. Abbott and Fairbanks, "Experiments on Plant Hybrids by Gregor Mendel," 410–11.
15. Abbott and Fairbanks, "Experiments on Plant Hybrids by Gregor Mendel," 411.
16. Abbott and Fairbanks, "Experiments on Plant Hybrids by Gregor Mendel," 411.
17. Abbott and Fairbanks, "Experiments on Plant Hybrids by Gregor Mendel," 411.
18. Abbott and Fairbanks, "Experiments on Plant Hybrids by Gregor Mendel," 411.
19. Abbott and Fairbanks, "Experiments on Plant Hybrids by Gregor Mendel," 411.
20. Abbott and Fairbanks, "Experiments on Plant Hybrids by Gregor Mendel," 411.
21. Abbott and Fairbanks, "Experiments on Plant Hybrids by Gregor Mendel," 412.
22. Abbott and Fairbanks, "Experiments on Plant Hybrids by Gregor Mendel," 412.
23. Abbott and Fairbanks, "Experiments on Plant Hybrids by Gregor Mendel," 414.
24. Abbott and Fairbanks, "Experiments on Plant Hybrids by Gregor Mendel," 413.

25. Abbott and Fairbanks, "Experiments on Plant Hybrids by Gregor Mendel," 413.
26. Abbott and Fairbanks, "Experiments on Plant Hybrids by Gregor Mendel," 413.
27. Abbott and Fairbanks, "Experiments on Plant Hybrids by Gregor Mendel," 414.

Chapter 6

1. National Academy of Sciences (USA), *Science and Creationism: A View from the National Academy of Sciences*, 2nd ed. (Washington, DC: National Academies Press, 1999), 2, https://www.nap.edu/catalog/6024/science-and-creationism-a-view-from-the-national-academy-of.
2. Scott Abbott and Daniel J. Fairbanks, "Experiments on Plant Hybrids by Gregor Mendel," *Genetics* 204 (2016): 416.
3. Abbott and Fairbanks, "Experiments on Plant Hybrids by Gregor Mendel," 416.
4. Abbott and Fairbanks, "Experiments on Plant Hybrids by Gregor Mendel," 416.
5. Abbott and Fairbanks, "Experiments on Plant Hybrids by Gregor Mendel," 416.
6. Abbott and Fairbanks, "Experiments on Plant Hybrids by Gregor Mendel," 416.
7. Abbott and Fairbanks, "Experiments on Plant Hybrids by Gregor Mendel," 417.
8. Abbott and Fairbanks, "Experiments on Plant Hybrids by Gregor Mendel," 417.
9. Jiří Sekerák, "Mendel in a Black Box," *Folia Mendeliana* 48, no. 2 (2012): 5.
10. Abbott and Fairbanks, "Experiments on Plant Hybrids by Gregor Mendel," 420.
11. Jiří Sekerák, "Mendel and Darwin," *Folia Mendeliana* 51, no. 2 (2015): 5; Jiří Sekerák, "At the Limits of (Our) Imagination: Did Mendel Really Fail to See the Importance of His Discovery for Darwin's Theory of Evolution?," *Folia Mendeliana* 55, no. 2 (2015): 19; Jiří Sekerák, *Anordnung: Mendel's Discovery of Inherited Information* (Brno: Moravian Museum, 2017).
12. Abbott and Fairbanks, "Experiments on Plant Hybrids by Gregor Mendel," 420.
13. Daniel J. Fairbanks, "Mendel and Darwin: Untangling a Persistent Enigma," *Heredity* 124 (2020): 263–73, https://www.nature.com/articles/s41437-019-0289-9.
14. Abbott and Fairbanks, "Experiments on Plant Hybrids by Gregor Mendel," 420.

15. Abbott and Fairbanks, "Experiments on Plant Hybrids by Gregor Mendel," 421.

16. Abbott and Fairbanks, "Experiments on Plant Hybrids by Gregor Mendel," 419.

17. Jiří Sekerák, "An Interpretation of Mendel's Discovery in the 21st Century," *Folia Mendeliana* 46 (2010): 23–40.

18. Ralf Dahm, "Friedrich Miescher and the Discovery of DNA," *Developmental Biology* 278 (2005): 274–88.

Chapter 7

1. Charles Darwin, *On the Origin of Species by Means of Natural Selection, or the Preservation of Favoured Races in the Struggle for Life*, 5th ed. (London: John Murray, London, 1869), 13.

2. Darwin, *Origin of Species*, 14.

3. Scott Abbot and Daniel J. Fairbanks, "Experiments on Plant Hybrids by Gregor Mendel," *Genetics* 204 (2016): 407.

4. Peter J. Van Dijk and T. H. Noel Ellis, "Mendel's Journey to Paris and London: Context and Significance for the Origin of Genetics," *Folia Mendeliana* 56 (2020): 5–34.

5. Daniel J. Fairbanks and Scott Abbott, "Alexander Makowsky's January 1865 Lecture 'On Darwin's Theory of Organic Creation': An English Translation with Commentary," *Folia Mendeliana* 55 (2019): 5–15.

6. Fairbanks and Abbott, "Alexander Makowsky's January 1865 Lecture," 8.

7. Abbott and Fairbanks, "Experiments on Plant Hybrids by Gregor Mendel," 408.

8. Charles Darwin, *On the Origin of Species by Means of Natural Selection, or the Preservation of Favoured Races in the Struggle for Life*, 3rd ed. (London: John Murray, 1861), 54–55.

9. Fairbanks and Abbott, "Alexander Makowsky's January 1865 Lecture," 11.

10. Abbott and Fairbanks, "Experiments on Plant Hybrids by Gregor Mendel," 422.

11. Fairbanks and Abbott, "Alexander Makowsky's January 1865 Lecture," 6.

12. Daniel J. Fairbanks and Bryce Rytting, "Mendelian Controversies: A Botanical and Historical Review," *American Journal of Botany* 88 (2001): 737–52, https://doi.org/10.2307/2657027.

13. Darwin, *Origin of Species*, 3rd ed., 296.

14. Daniel J. Fairbanks and Scott Abbott, "Darwin's Influence on Mendel: Evidence from a New Translation of Mendel's Paper," *Genetics* 204 (2016): 401–5, https://www.genetics.org/content/204/2/401.

15. Abbott and Fairbanks, "Experiments on Plant Hybrids by Gregor Mendel," 418–19.

16. Ronald A. Fisher, "Has Mendel's Work Been Rediscovered?" *Annals of Science* 1 (1936): 134.

17. Darwin, *Origin of Species*, 3rd ed., 7.

18. Fairbanks and Abbott, "Alexander Makowsky's January 1865 Lecture," 7–8.

19. Fairbanks and Abbott, "Darwin's Influence on Mendel," 401–5.

20. Darwin, *Origin of Species*, 3rd ed., 268.

21. Franz Unger, *Botanical Letters to a Friend*, trans. Eden Paul (Philadelphia: Lindsay and Blakiston, 1853), 92, https://www.biodiversitylibrary.org/item/215881#page/15/mode/1up.

22. Peter W. van der Pas, "A Note on the Reprints of Mendel's *Pisum* Paper," *Folia Mendeliana* 11 (1976): 53–54; Vítězslav Orel, *Gregor Mendel: The First Geneticist*, trans. Stephen Finn (Oxford: Oxford University Press, 1996), 276.

23. For example, Robin M. Henig, in *The Monk in the Garden: The Lost and Found Genius of Gregor Mendel, the Father of Genetics* (Boston: Houghton Mifflin Harcourt, 2000), claimed on page 143, "Another uncut reprint was found in the library of Charles Darwin, so Mendel must have sent him a copy, too. But even if Darwin had cut through the folds and try to read Mendel's paper, he might not have understood it."

24. Henig, *The Monk in the Garden*, 273.

25. Hermann Hoffmann, *Untersuchungen zur Bestimmung des Werthes von Species und Varietät: Ein Beitrag zur Kritik der Darwin'schen Hypothese* (Giessen: B. J. Ricker, 1869).

26. Robert C. Olby, *Origins of Mendelism*, 2nd ed. (Chicago: University of Chicago Press, 1985), 222.

27. Wilhelm O. Focke, *Die Pflanzen-Mischlinge; ein Beitrag zur Biologie der Gewächse* (Berlin: Gebrüder Borntraeger, 1881).

28. Charles Darwin, *The Formation of Vegetable Mould through the Action of Worms, with Observations on Their Habits* (London: John Murray, 1881).

29. Charles Darwin to George John Romanes, November 14, 1880, in Ethyl D. Romanes ed., *The Life and Letters of George John Romanes* (London: Longmans, Green, and Co., 1896), 105, biodiversitylibrary.org/bibliography/63198.

30. Olby, *Origins of Mendelism*, 228–29.

31. Charles Darwin, *The Variation of Animals and Plants under Domestication* (London: John Murray, 1868). Chapter 27 is titled "Provisional Hypothesis of Pangenesis" and is devoted entirely to it.

32. Abbot and Fairbanks, "Experiments on Plant Hybrids by Gregor Mendel," 409.

33. Darwin, *Origin of Species*, 3rd ed., 281.

34. Francis Darwin, ed., *The Life and Letters of Charles Darwin, Including an Autobiographical Chapter*, vol. 1 (London: John Murray, 1887), 93.

35. Henrietta E. Litchfield, ed., *Emma Darwin: Wife of Charles Darwin: A Century of Family Letters*, vol. 2 (Cambridge: Cambridge University Press, 1904), 230.

36. Daniel J. Fairbanks, "Mendel and Darwin: Untangling a Persistent Enigma," *Heredity* 124 (2020): 269, https://www.nature.com/articles/s41437-019-0289-9; Orel, *Gregor Mendel*, 194.

37. Darwin, *The Variation of Animals and Plants under Domestication*, 401

38. Curt Stern and Eva R. Sherwood, *The Origin of Genetics: A Mendel Source Book* (San Francisco: W. H. Freeman, 1966), 92.

39. Darwin, *The Variation of Animals and Plants under Domestication*, 363–64.

40. Stern and Sherwood, *The Origin of Genetics*, 93

41. Stern and Sherwood, *The Origin of Genetics*, 96.

42. Stern and Sherwood, *The Origin of Genetics*, 102.

43. Jiří Sekerák, "Mendel and Darwin," *Folia Mendeliana* 51, no. 2 (2015): 6; Jiří Sekerák, "At the Limits of (Our) Imagination: Did Mendel Really Fail to See the Importance of His Discovery for Darwin's Theory of Evolution?" *Folia Mendeliana* 55, no. 1 (2019): 19.

44. Hugo Iltis, *Life of Mendel*, trans. Eden Paul and Cedar Paul (New York: Hafner, 1966), 204; Fairbanks, "Mendel and Darwin," 271.

45. Orel, *Gregor Mendel*, 276.

46. Iltis, *Life of Mendel*, 205.

47. Fairbanks and Rytting, "Mendelian Controversies"; Fairbanks, "Mendel and Darwin."

Chapter 8

1. Peter F. Milovidov, "Gregor Mendel's Microscopic Preparations," *Folia Mendeliana* 3 (1968): 35–53.

2. Alfred H. Sturtevant, *A History of Genetics* (New York: Harper & Row, 1965), 12.

3. William R. Coleman, "Ferdinand Schindler Letters to William Bateson, 1902–1909," *Folia Mendeliana* 2 (1967): 10.

4. Vítězslav Orel, "Unknown Letters Relating to Mendel's State of Health," *Folia Mendeliana* 6 (1971): 270.

5. Hugo Iltis, *Life of Mendel*, trans. Eden Paul and Cedar Paul (New York: Hafner, 1966), 281.

6. Coleman, "Ferdinand Schindler Letters to William Bateson, 1902–1909," 9.

7. Daniel J. Fairbanks, "Mendel and Darwin: Untangling a Persistent Enigma," *Heredity* 124 (2020): 263–73, https://www.nature.com/articles/s41437-019-0289-9.

8. Siddhartha Mukherjee, *The Gene: An Intimate History* (New York: Scribner, 2016), 55.

9. Robin M. Henig, *The Monk in the Garden: The Lost and Found Genius of Gregor Mendel, the Father of Genetics* (Boston: Houghton Mifflin Harcourt, 2000), 161.

10. Peter van Dijk and T. H. Noel Ellis, "The Full Breadth of Mendel's Genetics," *Genetics* 204 (2016): 1327–36, https://www.genetics.org/node/435000.full.
11. Iltis, *Life of Mendel*, 191–92.
12. Curt Stern and Eva R. Sherwood, *The Origin of Genetics: A Mendel Source Book* (San Francisco: W. H. Freeman, 1966), 57.
13. Stern and Sherwood, *The Origin of Genetics*, 58.
14. Stern and Sherwood, *The Origin of Genetics*, 58–59.
15. Stern and Sherwood, *The Origin of Genetics*, 59.
16. Stern and Sherwood, *The Origin of Genetics*, 59.
17. Iltis, *Life of Mendel*, 191.
18. Iltis, *Life of Mendel*, 192.
19. Stern and Sherwood, *The Origin of Genetics*, 60.
20. Stern and Sherwood, *The Origin of Genetics*, 61.
21. Stern and Sherwood, *The Origin of Genetics*, 61.
22. Sturtevant, *A History of Genetics*, 11.
23. Henig, *The Monk in the Garden*, 159.
24. Mukherjee, *The Gene*, 55.
25. Iltis, *Life of Mendel*, 174.
26. Vítězslav Orel, *Gregor Mendel: The First Geneticist*, trans. Stephen Finn (Oxford: Oxford University Press, 1996), 184.
27. Carl von Nägeli and Albert Peter, *Die Hieracien Mittel-Europas* (Munich: Druck und Verlag von R. Oldenbourg, 1885).
28. This summary of reports to the Natural Science Society of Brünn on plant hybridization is from Orel, *Gregor Mendel*, 128–29.
29. Daniel J. Fairbanks and Scott Abbott, "Alexander Makowsky's January 1865 Lecture 'On Darwin's Theory of Organic Creation': An English Translation with Commentary," *Folia Mendeliana* 55 (2019): 8.
30. Stern and Sherwood, *The Origin of Genetics*, 50–51.
31. Carl Correns, *Gregor Mendels Briefe an Carl Nägeli 1866–1873* (Leipzig: G. G. Teubner, 1905), 211. An often-quoted translation of this passage is on page 71 of Stern and Sherwood, *The Origin of Genetics*, which they reproduced from a special supplement to the journal *Genetics* published in 1950. Unfortunately, this translation has been slightly embellished to make it more humorous. It uses the term "excess of avoirdupois" for "*Übergewichte*," which literally means "overweight" or "excess weight," and it adds the words "law of" when referring to "general gravitation" when those words are not in Mendel's writing. The translation here is my own, and it is a more accurate rendition of Mendel's original German.
32. Iltis, *Life of Mendel*, 277.
33. Stern and Sherwood, *The Origin of Genetics*, 86–87.
34. Stern and Sherwood, *The Origin of Genetics*, 73.
35. Van Dijk and Ellis, "The Full Breadth of Mendel's Genetics," 1327.
36. Stern and Sherwood, *The Origin of Genetics*, 55.

37. Stern and Sherwood, *The Origin of Genetics*, 93.

38. Robert C. Olby, *Origins of Mendelism*, 2nd ed. (Chicago: University of Chicago Press, 1985), 264.

39. Nägeli and Peter, *Die Hieracien Mittel-Europas*. The eight references to Mendel were found in an online search of the book available at http://books.google.com/books?id=DxwiAQAAMAAJ.

40. Olby, *Origins of Mendelism*, 264.

41. Stern and Sherwood, *The Origin of Genetics*, 97.

42. Stern and Sherwood, *The Origin of Genetics*, 79.

43. Gregor Mendel, "Bemerkungen zu der graphisch-tabelarischen Übersicht der meteorologische Verhältnisse von Brünn," *Verhandlungen des naturforschenden Vereines, Abhandlungen, Brünn* 1 (1863): 246–49; Gregor Mendel, "Meteorologische Beobachtungen aus Mähren und Schlesien für das Jahr 1863," *Verhandlungen des naturforschenden Vereines, Abhandlungen, Brünn* 2 (1864): 99–121; Gregor Mendel, "Meteorologische Beobachtungen aus Mähren für das Jahr 1864," *Verhandlungen des naturforschenden Vereines, Abhandlungen, Brünn* 3 (1865): 209–20; Gregor Mendel, "Meteorologische Beobachtungen aus Mähren und Schlesien für das Jahr 1865," *Verhandlungen des naturforschenden Vereines, Abhandlungen, Brünn* 4 (1866): 318–30; Gregor Mendel, "Meteorologische Beobachtungen aus Mähren und Schlesien für das Jahr 1866," *Verhandlungen des naturforschenden Vereines, Abhandlungen, Brünn* 5 (1867): 160–72; Gregor Mendel, "Meteorologische Beobachtungen aus Mähren und Schlesien für das Jahr 1869," *Verhandlungen des naturforschenden Vereines, Abhandlungen, Brünn* 8 (1870): 131–43. All of these references can be viewed online at https://www.biodiversitylibrary.org/bibliography/70769.

44. Translation by Scott Abbott.

45. Gregor Mendel, "Die Windhose vom 13. October 1870," *Verhandlungen des naturforschenden Vereines, Abhandlungen, Brünn* 9 (1871): 54–71, https://www.biodiversitylibrary.org/bibliography/70769.

46. Mendel, "Die Windhose," English translation by Scott Abbott and Daniel J. Fairbanks.

47. Gregor Mendel, "Die Grundlage der Wetterprognosen," *Mittheilungen* (Brünn) 56 (1879): 29–31.

48. Gregor Mendel, "Regenfall und Gewitter zu Brünn im Juni 1879," *Zeitschrift der Österreichischen Gesellschaft für Meteorologie* 14 (1879): 315–16; Gregor Mendel, "Gewitter in Brünn und Blansko am 15. August," *Zeitschrift der Österreichischen Gesellschaft für Meteorologie* 17 (1882): 407–8.

49. Anna Matalová, "The Beehouse of Gregor Mendel." *Folia Mendeliana* 17 (1982): 207–12.

50. Iltis, *Life of Mendel*, 213.

51. Iltis, *Life of Mendel*, 211.

52. Orel, *Gregor Mendel*, 240.

53. Morslav Vávra and Vítězslav Orel, "Hybridization of Pear Trees by Gregor Mendel," *Folia Mendeliana* 6 (1971): 190; Orel, *Gregor Mendel*, 224–25.

54. Gustav von Niessl, "Verzeichniss der Mitglieder (am Schlüsse des Jahres 1869)," *Verhandlungen des naturforschenden Vereines in Brünn* 8 (1870): XI–XXII. https://www.biodiversitylibrary.org/item/151433#page/14/mode/1up.

Chapter 9

1. Vítězslav Orel, *Gregor Mendel: The First Geneticist*, trans. Stephen Finn (Oxford: Oxford University Press, 1996), 212.

2. Hugo Iltis, *Life of Mendel*, trans. Eden Paul and Cedar Paul (New York: Hafner, 1966), 239.

3. Curt Stern and Eva R. Sherwood, *The Origin of Genetics: A Mendel Source Book* (San Francisco: W. H. Freeman, 1966), 79.

4. Gustav von Niessl, "Jahresversammlung am 21 December 1881," *Verhandlungen des naturforschenden Vereines in Brünn* 20 (1882): 45–48, https://www.biodiversitylibrary.org/bibliography/70769.

5. Adolar Zumkeller, "Recently Discovered Sermon Sketches of Gregor Mendel," *Folia Mendeliana* 6 (1971): 247–56.

6. Orel, *Gregor Mendel*, 261.

7. Anna Matalová, "Mendel's Confessions in the Ceiling Paintings in the Former Augustinian Monastery Prelacy," *Folia Mendeliana* 18 (1983): 273–76.

8. Anna Matalová, "Mendel's Experimental Plants Decorate the Augustinian Library Ceiling," *Folia Mendeliana* 20 (1985): 5–8.

9. Iltis, *Life of Mendel*, 242.

10. John Tyrrell, *Janáček: Years of a Life*, vol. 1, *(1854–1914): The Lonely Blackbird* (London: Faber and Faber, 2006).

11. Iltis, *Life of Mendel*, 242–43.

12. Orel, *Gregor Mendel*, 217.

13. Iltis, *Life of Mendel*, 264.

14. Iltis, *Life of Mendel*, 255.

15. Iltis, *Life of Mendel*, 256.

16. Iltis, *Life of Mendel*, 256–57.

17. Iltis, *Life of Mendel*, 257–58.

18. Iltis, *Life of Mendel*, 259.

19. Iltis, *Life of Mendel*, 261.

20. Iltis, *Life of Mendel*, 265.

21. Orel, *Gregor Mendel*, 261.

22. Orel, *Gregor Mendel*, 269.

23. Orel, *Gregor Mendel*, 269.

24. Iltis, *Life of Mendel*, 280, English translation by Daniel J. Fairbanks.

25. Anna Matalová, "Response to Mendel's Death in 1884," *Folia Mendeliana* 19 (1984): 217–21.

26. Matalová, "Response to Mendel's Death in 1884," 218.
27. Iltis, *Life of Mendel*, 281.

Chapter 10

1. Pam Hirsch and Mark McBeth, *Teacher Training at Cambridge: The Initiatives of Oscar Browning and Elizabeth Hughes* (London: Routledge, 2004), 192.
2. Christine Alexander, "Edith Rebecca Saunders," The Genetics Society, https://genetics.org.uk/news/filetodownload/edith-rebecca-saunders; Lawrence Hurst, "A Celebration of a Century of the Genetics Society (Founded by Edith Rebecca Saunders Ably Abetted by William Bateson)," *Folia Mendeliana* 55, no. 1 (2019): 23–28.
3. J. B. S. Haldane, "Miss E.R. Saunders," *Nature* 156 (1945): 385, https://www.nature.com/articles/156385b0.
4. Marsha L. Richmond, "Women in the Early History of Genetics: William Bateson and the Newnham College Mendelians, 1900–1910," *Isis* 92 (2001): 58.
5. Richmond, "Women in the Early History of Genetics," 64.
6. William Bateson and Edith Rebecca Saunders, "Experimental Studies in the Physiology of Heredity," *Reports to the Evolution Committee of the Royal Society* 1 (1902): 10–11.
7. Conrad Zirkle, "The Role of Liberty Hyde Bailey and Hugo de Vries in the Rediscovery of Mendelism," *Journal of the History of Biology* 1 (1968): 215.
8. Herbert F. Roberts, *Plant Hybridization before Mendel* (Princeton, NJ: Princeton University Press, 1929), 338, https://www.biodiversitylibrary.org/item/23454#page/9/mode/1up.
9. Hugo de Vries, "Sur la Loi de Disjonction des Hybrides," *Comptes Rendus de l'Académie des Sciences* 130 (1900): 845–47.
10. Carl Correns, "G. Mendel's Regel über das Verhalten der Nachkommenschaft der Rassenbastarde," *Berichte der deutschen botanischen Gesellschaft* 18 (1900): 158–68.
11. Curt Stern and Eva R. Sherwood, *The Origin of Genetics: A Mendel Source Book* (San Francisco: W. H. Freeman, 1966), 119–20.
12. Hugo de Vries, "Das Spaltzungsgesetz der Bastarde," *Berichte der deustchen botanischen Gesellschaft* 18 (1900): 83–90.
13. Erich von Tschermak, "Über künstliche Kreuzung bei *Pisum sativum*," *Berichte der deustchen botanischen Gesellschaft* 18 (1900): 232–39.
14. Beatrice Bateson, *William Bateson, F. R. S: His Essays & Addresses* (Cambridge: Cambridge University Press, 1928), 73.
15. Ida H. Stamhuis, Onno G. Meijer, and Erik J. A. Zevenhuisen, "Hugo de Vries on Heredity, 1889–1903: Statistics, Mendelian Laws, Pangenes, Mutations," *Isis* 90 (1999): 238–267; Onno G. Meijer, "Hugo de Vries No Mendelian!" *Annals of Science* 42 (1985): 189–232; Zirkle, "The Role of Liberty Hyde Bailey and Hugo de Vries in the Rediscovery of Mendelism," 205–18.

16. Hans-Jörg Reinberger, "When Did Carl Correns Read Gregor Mendel's Paper? A Research Note," *Isis* 86 (1995): 612–16.

17. Stern and Sherwood, *The Origin of Genetics*, x–xi; Curt Stern and Evelyn Stern, "A Note on the 'Three Rediscoverers' of Mendelism," *Folia Mendeliana* 13 (1978): 237–40.

18. Robert C. Olby, "William Bateson's Introduction of Mendelism to England: A Reassessment," *British Journal for the History of Science* 20 (1987): 399–420.

19. Bateson and Saunders, "Experimental Studies in the Physiology of Heredity."

20. Richmond, "Women in the Early History of Genetics," 69–71.

21. Richmond, "Women in the Early History of Genetics," 71.

22. Charles C. Hurst, "Mendel's Principles Applied to Wheat Hybrids," *Journal of the Royal Horticultural Society* 27 (1903): 876–93.

23. William B. Provine, *The Origins of Theoretical Population Genetics* (Chicago: University of Chicago Press, 1971), 68.

24. Scott Abbott and Daniel J. Fairbanks, "Experiments on Plant Hybrids by Gregor Mendel," *Genetics* 204 (2016): 408.

25. Abbott and Fairbanks, "Experiments on Plant Hybrids by Gregor Mendel," 414.

26. Robert C. Olby, *Origins of Mendelism*, 2nd ed. (Chicago: University of Chicago Press, 1985), 115.

27. Francis Galton, *Natural Inheritance* (London: Macmillan & Co., 1889), 12.

28. W. F. R. Weldon, "Mendel's Law of Alternative Inheritance in Peas," *Biometrika* 1 (1902): 228–54.

29. Weldon, "Mendel's Law of Alternative Inheritance in Peas," 232–35.

30. Weldon, "Mendel's Law of Alternative Inheritance in Peas," 252.

31. William Bateson, *Mendel's Principles of Inheritance: A Defence* (Cambridge: Cambridge University Press, 1902).

32. Bateson, *Mendel's Principles of Inheritance*, vi.

33. Bateson, *Mendel's Principles of Inheritance*, 208.

34. George Udny Yule, "Mendel's Laws and Their Probable Relations to Intra-Racial Heredity," *New Phytologist* 1 (1902): 194.

35. Yule, "Mendel's Laws and Their Probable Relations to Intra-Racial Heredity," 236.

36. Archibald E. Garrod, "The Incidence of Alkaptonuria: A Study in Chemical Individuality," *Lancet* 160 (1902): 1616–20.

37. Provine, *The Origins of Theoretical Population Genetics*, 74.

38. W. F. R. Weldon, "Mr. Bateson's Revisions of Mendel's Theory of Heredity," *Biometrika* 2 (1903): 286.

39. Weldon, "Mr. Bateson's Revisions of Mendel's Theory of Heredity," 298.

40. Provine, *The Origins of Theoretical Population Genetics*, 79.

41. Reginald C. Punnett, "Early Days of Genetics," *Heredity* 4 (1950): 7–8.

42. Wilhelm Johannsen, *Elemente der exakten Erblichkeitslehre* (Jena: Verlag von Gustav Fischer, 1909).

43. Hermann Nillson-Ehle, "Kreutzungsuntersuchungen an Hafer und Weizen," *Lunds Universitets Årsskrift* (1909).

44. Edward M. East, "A Mendelian Interpretation of Variation That Is Apparently Continuous," *American Naturalist* 44 (1910): 65–82.

45. Edward M. East, "Studies on Size Inheritance in Nicotiana," *Genetics* 1 (1916): 164–76.

46. William Wilkes, ed., *Report of the Third International Conference 1906 on Genetics* (London: Royal Horticultural Society, 1906).

47. Wilkes, *Report of the Third International Conference 1906 on Genetics*.

48. Wilkes, *Report of the Third International Conference 1906 on Genetics*, 70.

Epilogue

1. Anna Matalová and Eva Matalová, "Czech Centre Marks Mendel Anniversary," *Nature* 518 (2015): 303, https://www.nature.com/articles/518303e.

2. Vítězslav Orel, *Gregor Mendel: The First Geneticist*, trans. Stephen Finn (Oxford: Oxford University Press, 1996), 314.

3. Vítězslav Orel, "Jaroslav Kříženecký (1896–1964), Tragic Victim of Lysenkoism in Czechoslovakia," *Quarterly Review of Biology* 67 (1992): 492.

4. Orel, "Jaroslav Kříženecký (1896–1964), Tragic Victim of Lysenkoism in Czechoslovakia," 487.

5. I. Michael Lerner, foreword to Zhores A. Medvedev, *The Rise and Fall of T. D. Lysenko*, trans. I. Michael Lerner (New York: Columbia University Press, 1969), v.

6. Julian Huxley, *Soviet Genetics and World Science* (London: Chatto and Windus, 1949), 17.

7. Medvedev, *The Rise and Fall of T. D. Lysenko*, 54–55.

8. Medvedev, *The Rise and Fall of T. D. Lysenko*, 58.

9. Medvedev, *The Rise and Fall of T. D. Lysenko*, 72.

10. Gary P. Nabhan, *Where Our Food Comes From: Retracing Nikolay Vavilov's Quest to End Famine* (Washington, DC: Island Press, 2009), 10.

11. Orel, "Jaroslav Kříženecký (1896–1964), Tragic Victim of Lysenkoism in Czechoslovakia," 487–494.

12. Orel, *Gregor Mendel*, 314.

13. Anna Matalová and Jiří Sekerák, *Genetics behind the Iron Curtain: Its Repudiation and Reinstitutionalisation in Czechoslovakia* (Brno: Moravian Museum, 2004).

14. Orel, "Jaroslav Kříženecký (1896–1964), Tragic Victim of Lysenkoism in Czechoslovakia," 491.

15. Medvedev, *The Rise and Fall of T. D. Lysenko*, 204–5.

16. Orel, "Jaroslav Kříženecký (1896–1964), Tragic Victim of Lysenkoism in Czechoslovakia," 493.

17. Orel, "Jaroslav Kříženecký (1896–1964), Tragic Victim of Lysenkoism in Czechoslovakia," 492.

Appendix

1. One variety has a beautiful brown-red pod colour that transforms into violet and blue at the time of ripening. The experiment with this character was begun only during the past year.

2. With Pisum it is shown without doubt that there must be a complete union of the elements of both fertilising cells for the formation of the new embryo. How could one otherwise explain that among the progeny of hybrids both original forms reappear in equal number and with all their peculiarities. If the influence of the germ cell on the pollen cell were only external, if it were given only the role of a nurse, then the result of every artificial fertilisation could only be that the developed hybrid was exclusively like the pollen plant or was very similar to it. In no manner have experiments until now confirmed that. Fundamental evidence for the complete union of the contents of both cells lies in the universally confirmed experience that it is unimportant for the form of the hybrid which of the original forms was the seed or the pollen plant.

Bibliography

Abbott, Scott, and Daniel J. Fairbanks. "Experiments on Plant Hybrids by Gregor Mendel." *Genetics* 204 (2016): 407–22.

Alexander, Christine. "Edith Rebecca Saunders." Genetics Society. https://genetics.org.uk/news/filetodownload/edith-rebecca-saunders.

Auspitz, Josef, to Gregor Mendel, Brünn, March 13, 1858. Collection of the Mendelianum of the Moravian Museum, Brno, Czech Republic.

Bateson, Beatrice. *William Bateson, F. R. S: His Essays & Addresses*. Cambridge: Cambridge University Press, 1928.

Bateson, William. *Mendel's Principles of Inheritance: A Defence*. Cambridge: Cambridge University Press, 1902.

Bateson, William, and E. Rebecca Saunders. "Experimental Studies in the Physiology of Heredity." *Reports to the Evolution Committee of the Royal Society* 1 (1902): 3–160.

Bickel, Lennard. *Facing Starvation: Norman Borlaug and the Fight against Hunger*. Pleasantville, NY: Reader's Digest Press, 1974.

Bishop, B. E. "Mendel's Opposition to Evolution and to Darwin." *Journal of Heredity* 87 (1996): 205–13.

Callender, L. A. "Gregor Mendel: An Opponent of Descent with Modification." *History of Science* 26 (1988): 41–75.

Christies of London. "Important Scientific Books from the Collection of Peter and Margarethe Braune, sale 17700, lot 307," 2019. https://www.christies.com/lotfinder/Lot/mendel-johann-gregor-1822-1884-versuche-uber-pflanzen-hybriden-6216731-details.aspx.

Coleman, William R. "Ferdinand Schindler Letters to William Bateson, 1902–1909." *Folia Mendeliana* 2 (1967): 9–17.

Correns, Carl. "G. Mendel's Regel über das Verhalten der Nachkommenschaft der Rassenbastarde." *Berichte der deutschen botanischen Gesellschaft* 18 (1900): 158–68.

———. *Gregor Mendels Briefe an Carl Nägeli 1866–1873*. Leipzig: G. G. Teubner, 1905.

Dahm, Ralf. "Friedrich Miescher and the Discovery of DNA." *Developmental Biology* 278 (2005): 274–88.

Darbishire, Arthur D. *Breeding and the Mendelian Discovery*. London: Cassell and Company, 1911.

Darwin, Charles. *On the Origin of Species by Means of Natural Selection, or the Preservation of Favoured Races in the Struggle for Life*. London: John Murray, 1859.

———. *On the Origin of Species by Means of Natural Selection, or the Preservation of Favoured Races in the Struggle for Life*, 3rd ed. London: John Murray, 1861.

———. *The Variation of Animals and Plants under Domestication*. London: John Murray, 1868.

———. *On the Origin of Species by Means of Natural Selection, or the Preservation of Favoured Races in the Struggle for Life*, 5th ed. London: John Murray, 1869.

———. *The Formation of Vegetable Mould through the Action of Worms, with Observations on Their Habits*. London: John Murray, 1881.

Darwin, Francis, ed. *The Life and Letters of Charles Darwin, Including an Autobiographical Chapter*. Vol. 1. London: John Murray, 1887.

De Vries, Hugo. "Das Spaltzungsgesetz der Bastarde." *Berichte der deustchen botanischen Gesellschaft* 18 (1900): 83–90.

———. "Sur la Loi de Disjonction des Hybrides." *Comptes Rendus de l'Académie des Sciences* 130 (1900): 845–47.

East, Edward M. "A Mendelian Interpretation of Variation That Is Apparently Continuous." *American Naturalist* 44 (1910): 65–82.

———. "Studies on Size Inheritance in Nicotiana." *Genetics* 1 (1916): 164–76.

Eisley, Loren. *Darwin's Century: Evolution and the Men Who Discovered It*. London: Scientific Book Guild, 1959.

Fairbanks, Daniel J. "Mendel and Darwin: Untangling a Persistent Enigma." *Heredity* 124 (2020): 263–73. https://www.nature.com/articles/s41437-019-0289-9.

Fairbanks, Daniel J., and Scott Abbott. "Darwin's Influence on Mendel: Evidence from a New Translation of Mendel's Paper." *Genetics* 204 (2016): 401–5. https://www.genetics.org/content/204/2/401.

———. "Alexander Makowsky's January 1865 Lecture 'On Darwin's Theory of Organic Creation': An English Translation with Commentary." *Folia Mendeliana* 55 (2019): 5–15.

Fairbanks, Daniel J., and Bryce Rytting. "Mendelian Controversies: A Botanical and Historical Review." *American Journal of Botany* 88 (2001): 737–52. https://doi.org/10.2307/2657027.

Farmer, James L., and Daniel J. Fairbanks. "Interaction of the *bw* and *w* loci in *Drosophila melanogaster*." *Genetics* 107 (1984): s30.

Fisher, Ronald A. "Has Mendel's Work Been Rediscovered?" *Annals of Science* 1 (1936): 115–37.

Focke, Wilhelm O. *Die Pflanzen-Mischlinge; ein Beitrag zur Biologie der Gewächse.* Berlin: Gebrüder Borntraeger, 1881.

Franklin, Allan., Anthony W. F. Edwards, Daniel J. Fairbanks, Daniel L. Hartl, and Teddy Seidenfeld. *Ending the Mendel–Fisher Controversy.* Pittsburgh, PA: University of Pittsburgh Press, 2008.

Galton, Francis. *Natural Inheritance.* London: Macmillan & Co., 1889.

Garrod, Archibald E. "The Incidence of Alkaptonuria: A Study in Chemical Individuality." *Lancet* 160 (1902): 1616–20.

Gärtner, Carl F. *Versuche und Beobachtungen über die Bastarderzeugung im Pflanzenreich.* Stuttgart: K. F. Herring, 1849.

Gliboff, Sander. "Evolution, Revolution, and Reform in Vienna: Franz Unger's Ideas on Descent and Their Post-1848 Reception." *History of Science* 37 (1998): 217–35.

Haldane, J. B. S. "Miss E.R. Saunders." *Nature* 156 (1945): 385. https://www.nature.com/articles/156385b0.

Hartl, Daniel L., and Daniel J. Fairbanks. "Mud Sticks: On the Alleged Falsification of Mendel's Data." *Genetics* 175 (2007): 975–79. https://www.genetics.org/content/175/3/975.

Henig, Robin M. *The Monk in the Garden: The Lost and Found Genius of Gregor Mendel, the Father of Genetics.* Boston: Houghton Mifflin Harcourt, 2000.

Hirsch, Pam, and Mark McBeth. *Teacher Training at Cambridge: The Initiatives of Oscar Browning and Elizabeth Hughes.* London: Routledge, 2004.

Hoffmann, Hermann. *Untersuchungen zur Bestimmung des Werthes von Species und Varietät: ein Beitrag zur Kritik der Darwin'schen Hypothese.* Giessen: B. J. Ricker, 1869.

Hurst, Charles C. "Mendel's Principles Applied to Wheat Hybrids." *Journal of the Royal Horticultural Society* 27 (1903): 876–93.

Hurst, Lawrence. "A Celebration of a Century of the Genetics Society (Founded by Edith Rebecca Saunders Ably Abetted by William Bateson)." *Folia Mendeliana* 55, no. 1 (2019): 23–28.

Huxley, Julian. *Soviet Genetics and World Science.* London: Chatto and Windus, 1949.

Huxley, Thomas H. "Science and 'Church Policy.'" *Reader,* December 31, 1864.

Iltis, Anne. "Gregor Mendel's Autobiography." *Journal of Heredity* 45 (1954): 231–34.

Iltis, Hugo. *Gregor Johann Mendel: Leben, Werk und Wirkung.* Berlin: Springer, 1924. https://www.google.com/books/edition/Gregor_Johann_Mendel/AD0MAAAAMAAJ?hl.

———. *Life of Mendel.* Translated by Eden Paul and Cedar Paul. New York: Hafner, 1966.

Johannsen, Wilhelm. *Elemente der exakten Erblichkeitslehre.* Jena: Verlag von Gustav Fischer, 1909.

Keller, Evelyn F. *A Feeling for the Organism: The Life and Work of Barbara McClintock*. San Francisco: W. H. Freeman, 1983.

Klein, Jan, and Norman Klein. *Solitude of a Humble Genius—Gregor Johann Mendel*. Vol. 1. *Formative Years*. Berlin: Springer, 2013.

Křiženecký, I. Jaroslav. "Mendels zweite erfolglose Lehramtsprüfung im Jahre 1856." *Sudhoffs Archiv für Geschichte der Medizin und der Naturwissenschaften* 47 (1963): 305–10.

Lerner, I. Michael. Foreword to *The Rise and Fall of T. D. Lysenko*, by Zhores A. Medvedev. Translated by I. Michael Lerner. New York: Columbia University Press, 1969.

Litchfield, Henrietta E., ed. *Emma Darwin: Wife of Charles Darwin: A Century of Family Letters*. Vol. 2. Cambridge: Cambridge University Press, 1904.

Mährischer Correspondent (Anonymous). "Wetter vom 11 Jänner." *Mährischer Correspondent*, January 13, 1865. http://www.digitalniknihovna.cz/mzk/view/uuid:04530090-64ba-11e3-8c6a-005056825209?page=uuid:08aa19f0-6710-11e3-8387-001018b5eb5c.

Matalová, Anna. "A Monument to F. M. Klácel (1809–1882) in the Vicinity of the Mendel Statue in Brno." *Folia Mendeliana* 14 (1973): 251–63.

———. "The Beehouse of Gregor Mendel." *Folia Mendeliana* 17 (1982): 207–12.

———. "Mendel's Confessions in the Ceiling Paintings in the Former Augustinian Monastery Prelacy." *Folia Mendeliana* 18 (1983): 273–76.

———. "Response to Mendel's Death in 1884." *Folia Mendeliana* 19 (1984): 217–21.

———. "Mendel's Experimental Plants Decorate the Augustinian Library Ceiling." *Folia Mendeliana* 20 (1985): 5–8.

Matalová, Anna, and Eva Matalová. "Czech Centre Marks Mendel Anniversary." *Nature* 518 (2015): 303. https://www.nature.com/articles/518303e.

Matalová, Anna, and Jiří Sekerák. *Genetics behind the Iron Curtain: Its Repudiation and Reinstitutionalisation in Czechoslovakia*. Brno: Moravian Museum, 2004.

Medvedev, Zhores A. *The Rise and Fall of T. D. Lysenko*. Translated by I. Michael Lerner. New York: Columbia University Press, 1969.

Meijer, Onno G. "Hugo de Vries No Mendelian!" *Annals of Science* 42 (1985): 189–232.

Mendel, Gregor. "Über Verwüstung am Gartenrettig durch Raupen (*Botys margaritalis*)." *Verhandlungen des zoologisch-botanischen Vereins in Wien* 3 (1853): 116–18. https://www.biodiversitylibrary.org/bibliography/16346.

———. "Beschreibung des sogenannten Erbsenkäfers, *Bruchus pisi*, Mitgeteilt von V. Kollar." *Verhandlungen des zoologisch-botanischen Vereins in Wien* 4 (1854): 27–30. https://www.biodiversitylibrary.org/bibliography/16346.

———. "Bemerkungen zu der graphisch-tabelarischen Übersicht der meteorologische Verhältnisse von Brünn." *Verhandlungen des naturforschenden Vereines in Brünn, Abhandlungen* 1 (1863): 246–49. https://www.biodiversitylibrary.org/bibliography/70769.

———. "Meteorologische Beobachtungen aus Mähren und Schlesien für das Jahr 1863." *Verhandlungen des naturforschenden Vereines in Brünn, Abhandlungen* 2 (1864): 99–121. https://www.biodiversitylibrary.org/bibliography/70769.

———. "Meteorologische Beobachtungen aus Mähren für das Jahr 1864." *Verhandlungen des naturforschenden Vereines in Brünn, Abhandlungen* 3 (1865): 209–20. https://www.biodiversitylibrary.org/bibliography/70769.

———. "Meteorologische Beobachtungen aus Mähren und Schlesien für das Jahr 1865." *Verhandlungen des naturforschenden Vereines in Brünn, Abhandlungen* 4 (1866): 318–30. https://www.biodiversitylibrary.org/bibliography/70769.

———. "Versuche über Pflanzen-Hybriden." *Verhandlungen des naturforschenden Vereines in Brünn, Abhandlungen* 4 (1866): 3–47. https://www.biodiversitylibrary.org/bibliography/70769.

———. "Meteorologische Beobachtungen aus Mähren und Schlesien für das Jahr 1866." *Verhandlungen des naturforschenden Vereines in Brünn, Abhandlungen* 5 (1867): 160–72. https://www.biodiversitylibrary.org/bibliography/70769.

———. "Meteorologische Beobachtungen aus Mähren und Schlesien für das Jahr 1869." *Verhandlungen des naturforschenden Vereines in Brünn, Abhandlungen* 8 (1870): 131–43. https://www.biodiversitylibrary.org/bibliography/70769.

———. "Über einige aus künstlicher Befruchtung gewonnenen Hieracium-Bastarde." *Verhandlungen des naturforschenden Vereines in Brünn* 8 (1870): 26–31. https://www.biodiversitylibrary.org/bibliography/70769.

———. "Die Windhose vom 13. October 1870." *Verhandlungen des naturforschenden Vereines in Brünn, Abhandlungen* 9 (1871): 54–71. https://www.biodiversitylibrary.org/bibliography/70769.

———. "Die Grundlage der Wetterprognosen." *Mittheilungen* (Brünn) 56 (1879): 29–31.

———. "Regenfall und Gewitter zu Brünn im Juni 1879." *Zeitschrift der Österreichischen Gesellschaft für Meteorologie* 14 (1879): 315–16.

———. "Gewitter in Brünn und Blansko am 15. August." *Zeitschrift der Österreichischen Gesellschaft für Meteorologie* 17 (1882): 407–8.

Milovidov, Peter F. "Gregor Mendel's Microscopic Preparations." *Folia Mendeliana* 3 (1968): 35–53.

"Monats-Versammlung des naturforschenden Vereins in Brünn am 8 März 1865." *Brünner Zeitung*, no. 65, March 20, 1865. http://www.digitalniknihovna.cz/mzk/view/uuid:e32cb9e0-f061-11e3-a012-005056825209?page=uuid:59ba8440-f7f9-11e3-8232-5ef3fc9ae867.

Mukherjee, Siddhartha. *The Gene: An Intimate History*. New York: Scribner, 2016.

Nabhan, Gary P. *Where Our Food Comes From: Retracing Nikolay Vavilov's Quest to End Famine*. Washington, DC: Island Press, 2009.

Nägeli, Carl von, and Albert Peter. *Die Hieracien Mittel-Europas*. Munich: Druck und Verlag von R. Oldenbourg, 1885.

National Academy of Sciences (USA). *Science and Creationism: A View from the National Academy of Sciences*, 2nd ed. Washington, DC: National Academies Press, 1999.

https://www.nap.edu/catalog/6024/science-and-creationism-a-view-from-the-national-academy-of.

Niessl, Gustav von. "Jahresversammlung am 21 December 1881." *Verhandlungen des naturforschenden Vereines in Brünn* 20 (1882): 45–48. https://www.biodiversitylibrary.org/bibliography/70769.

———. "Verzeichniss der Mitglieder (am Schlüsse des Jahres 1869)." *Verhandlungen des naturforschenden Vereines in Brünn* 8 (1870): xi–xxii. https://www.biodiversitylibrary.org/bibliography/70769.

Nillson-Ehle, Herman. *Kreutzungsuntersuchungen an Hafer und Weizen.* Lund: Lunds Universitets Årsskrift, 1909.

Olby, Robert C. "Mendel no Mendelian?" *History of Science* 17 (1979): 53–72.

———. *Origins of Mendelism.* 2nd ed. Chicago: University of Chicago Press, 1985.

Orel, Vítězslav. "Unknown Letters Relating to Mendel's State of Health." *Folia Mendeliana* 6 (1971): 265–70.

———. "Mendel and New Scientific Ideas at the Vienna University." *Folia Mendeliana* 7 (1972): 27–36.

———. "Jaroslav Kříženecký (1896–1964), Tragic Victim of Lysenkoism in Czechoslovakia." *Quarterly Review of Biology* 67 (1992): 487–94.

———. *Gregor Mendel: The First Geneticist.* Translated by Stephen Finn. Oxford: Oxford University Press, 1996.

Orel, Vítězslav, Gerhard Czihak, and Hans Wieseneder. "Mendel's Examination Paper on the Geological Formation of the Earth of 1850." *Folia Mendeliana* 18 (1983): 227–72.

Orel, Vítězslav, and Antonín Verbík. "Mendel's Involvement in the Plea for Freedom of Teaching in the Revolutionary Year of 1848." *Folia Mendeliana* 19 (1984): 223–33.

Paigen, Kenneth. "One Hundred Years of Mouse Genetics: An Intellectual History. I. The Classical Period (1902–1980)." *Genetics* 163 (2003): 1–7. https://www.genetics.org/content/163/1/1.

Peaslee, Margaret H., and Vítězslav Orel. "The Evolutionary Ideas of F. M. (Ladimir) Klacel, Teacher of Gregor Mendel." *Biomedical Papers of the Medical Faculty of the University of Palacky, Olomouc, Czech Republic* 151 (2007): 151–56.

Phifer-Rixey, Megan, and Michael W. Nachman. "Insights into Mammalian Biology from the Wild House Mouse *Mus musculus*." *eLife* 4 (2015): e05959. https://elifesciences.org/articles/05959.

Provine, William B. *The Origins of Theoretical Population Genetics.* Chicago: University of Chicago Press, 1971.

Punnett, Reginald C. "Early Days of Genetics." *Heredity* 4 (1950): 1–10.

Reese, R. Allan. "Darwin, Mendel, and the Evolution of Evolution." *Significance* 6 (2009): 127–29. https://rss.onlinelibrary.wiley.com/doi/full/10.1111/j.1740-9713.2009.00377.x.

Reinberger, Hans-Jörg. "When Did Carl Correns Read Gregor Mendel's Paper? A Research Note." *Isis* 86 (1995): 612–16.

Richmond, Marsha L. "Women in the Early History of Genetics: William Bateson and the Newnham College Mendelians, 1900–1910." *Isis* 92 (2001): 55–90.

Roberts, Herbert F. *Plant Hybridization before Mendel*. Princeton, NJ: Princeton University Press, 1929. https://www.biodiversitylibrary.org/item/23454#page/9/mode/1up.

Romanes, Ethel D., ed. *The Life and Letters of George John Romanes*. London: Longmans, Green, and Co., 1896. https://www.biodiversitylibrary.org/bibliography/63198.

Sekerák, Jiří. "An Interpretation of Mendel's Discovery in the 21st Century." *Folia Mendeliana* 46 (2010): 23–40.

———. "Mendel in a Black Box." *Folia Mendeliana* 48, no. 2 (2012): 5–36.

———. "Mendel and Darwin." *Folia Mendeliana* 51, no. 2 (2015): 5–9.

———. *Anordnung: Mendel's Discovery of Inherited Information*. Brno: Moravian Museum, 2017.

———. "Mendel's Scientific Society in Brno." *Folia Mendeliana* 54 (2018): 5–10.

———. "At the Limits of (Our) Imagination: Did Mendel Really Fail to See the Importance of His Discovery for Darwin's Theory of Evolution?" *Folia Mendeliana* 55, no. 1 (2019): 17–21.

Stamhuis, Ida H., Onno G. Meiher, and Erik J. A. Zevenhuisen. "Hugo de Vries on Heredity, 1889–1903: Statistics, Mendelian Laws, Pangenes, Mutations." *Isis* 90 (1999): 238–67.

Stern, Curt, and Eva R. Sherwood. *The Origin of Genetics: A Mendel Source Book*. San Francisco: W. H. Freeman, 1966.

Stern, Curt, and Evelyn Stern. "A Note on the 'Three Rediscoverers' of Mendelism." *Folia Mendeliana* 13 (1978): 237–40.

Sturtevant, Alfred H. *A History of Genetics*. New York: Harper & Row, 1965.

Toperczer, Max. "Liznar, Josef (1852–1932), Geophysiker und Meteorologe." *Österreichishes Biographisches Lexikon* 5 (1971): 254. https://www.biographien.ac.at/oebl/oebl_L/Liznar_Josef_1852_1932.xml.

Tschermak, Erich. "Über künstliche Kreuzung bei *Pisum sativum*." *Berichte der deustchen botanischen Gesellschaft* 18 (1900): 232–39. https://www.biodiversitylibrary.org/item/132842#page/296/mode/1up.

Tyrrell, John. *Janáček: Years of a Life—Volume 1 (1854–1914): The Lonely Blackbird*. London: Faber and Faber, 2006.

Unger, Franz. *Die Urwelt in ihren verschiedenen Bildungsperioden*. Vienna: Fr. Beck, 1851. https://gdz.sub.uni-goettingen.de/id/PPN782695469.

———. *Botanische Briefe*. Vienna: Verlag von Carl Gerold & Sohn, 1852. https://www.biodiversitylibrary.org/item/47372#page/11/mode/1up.

———. *Botanical Letters to a Friend*. Translated by Eden Paul. Philadelphia: Lindsay and Blakiston, 1853. https://www.biodiversitylibrary.org/item/215881#page/15/mode/1up.

Van der Pas, Peter W. "The Date of Gregor Mendel's Birth." *Folia Mendeliana* 7 (1972): 7–12.

Van der Weyden, Louise, Jaqueline K. White, David J. Adams, and Darren W. Logan. "The Mouse Genetic Toolkit: Revealing Function and Mechanism." *Genome Biology* 12 (2011): 224. https://genomebiology.biomedcentral.com/articles/10.1186/gb-2011-12-6-224.

Van Dijk, Peter J. "Gregor Mendel's Meeting with Pope Pius IX: The Truth in the Story." *Folia Mendeliana* 56 (2020): 35–50.

Van Dijk, Peter J., and T. H. Noel Ellis. "The Full Breadth of Mendel's Genetics." *Genetics* 204 (2016): 1327–36. https://www.genetics.org/node/435000.full.

———. "Mendel's Journey to Paris and London: Context and Significance for the Origin of Genetics." *Folia Mendeliana* 56 (2020): 5–34.

Vávra, Morslav, and Vítězslav Orel. "Hybridization of Pear Trees by Gregor Mendel." *Folia Mendeliana* 6 (1971): 189–91.

Vollmann, Johann, and Anna Matalová. "Echoes of Mendel's Life and Work in Newspapers between the Years 1850–1884." *Folia Mendeliana* 52, no. 1 (2016): 21–37.

Weir, John A. "Agassiz, Mendel, and Heredity." *Journal of the History of Biology* 1 (1968): 179–203.

Weldon, W. F. R. "Mendel's Law of Alternative Inheritance in Peas." *Biometrika* 1 (1902): 228–254.

———. "Mr. Bateson's Revisions of Mendel's Theory of Heredity." *Biometrika* 2 (1903): 286–98.

Wilkes, William, ed. *Report of the Third International Conference 1906 on Genetics.* London: Royal Horticultural Society, 1906. https://www.biodiversitylibrary.org/item/206746#page/5/mode/1up.

Yule, G. Udny. "Mendel's Laws and Their Probable Relations to Intra-Racial Heredity." *New Phytologist* 1 (1902): 193–238.

Zhang, Hui, Wen Chen, and Kun Sun. "Mendelism: New Insights from Gregor Mendel's Lectures in Brno." *Genetics* 207 (2017): 1–8. https://www.genetics.org/content/207/1/1.

Zirkle, Conrad. "The Role of Liberty Hyde Bailey and Hugo de Vries in the Rediscovery of Mendelism." *Journal of the History of Biology* 1 (1968): 205–18.

Zumkeller, Adolar. "Recently Discovered Sermon Sketches of Gregor Mendel." *Folia Mendeliana* 6 (1971): 247–56.

Index

Note: Page numbers in *italics* refer to drawings and photos.